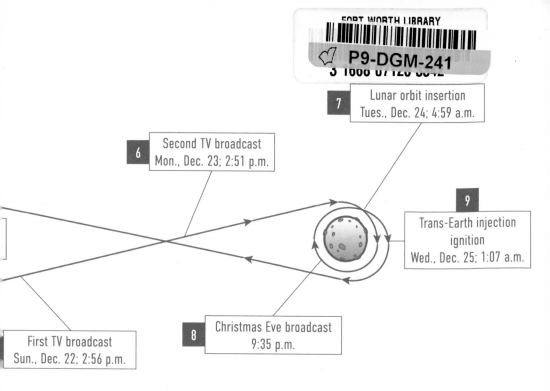

7 Lunar orbit insertion
Tues., Dec. 24; 4:59 a.m.

6 Second TV broadcast
Mon., Dec. 23; 2:51 p.m.

9 Trans-Earth injection
ignition
Wed., Dec. 25; 1:07 a.m.

8 Christmas Eve broadcast
9:35 p.m.

First TV broadcast
Sun., Dec. 22; 2:56 p.m.

APOLLO 8 COMMAND MODULE/SERVICE MODULE

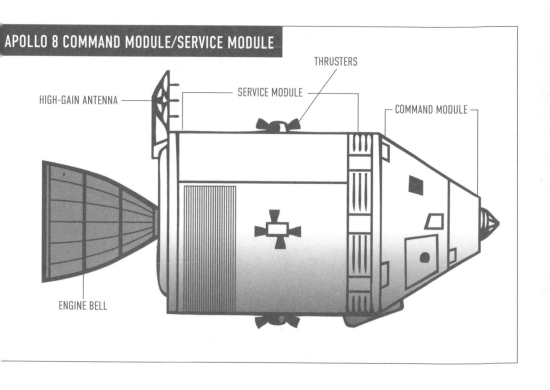

HIGH-GAIN ANTENNA

THRUSTERS

SERVICE MODULE

COMMAND MODULE

ENGINE BELL

Also by Jeffrey Kluger

APOLLO

8

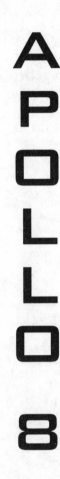

APOLLO 8

The Thrilling Story of the First Mission to the Moon

JEFFREY KLUGER

Henry Holt and Company   New York

Henry Holt and Company

Publishers since 1866
175 Fifth Avenue
New York, New York 10010
www.henryholt.com

Henry Holt ® and ® are registered trademarks of
Macmillan Publishing Group, LLC.

Library of Congress Cataloging-in-Publication Data
Names: Kluger, Jeffrey, author.
Title: Apollo 8 : the thrilling story of the first mission to the Moon / Jeffrey Kluger.
Description: New York : Henry Holt and Co., 2017. | Includes bibliographical
   references and index.
Identifiers: LCCN 2016046157 | ISBN 9781627798327 (hardback) | ISBN
   9781627798310 (electronic book)
Subjects: LCSH: Project Apollo (U.S.) | Space flight to the moon. | BISAC:
   HISTORY / Expeditions & Discoveries. | HISTORY / United States / 20th
   Century. | TECHNOLOGY & ENGINEERING / Aeronautics & Astronautics.
Classification: LCC TL789.8.U6 A54325 2017 | DDC 629.45/4—dc23
LC record available at https://lccn.loc.gov/2016046157

Our books may be purchased in bulk for promotional, educational, or business
use. Please contact your local bookseller or the Macmillan Corporate and
Premium Sales Department at (800) 221-7945, extension 5442, or by e-mail at
MacmillanSpecialMarkets@macmillan.com.

First Edition 2017

Endpaper illustrations by Lisa Marie Brennan, lisa@saber.net

Designed by Meryl Sussman Levavi

Printed in the United States of America

1 3 5 7 9 10 8 6 4 2

With love to Alejandra, Elisa, and Paloma—
for the bright sunlight
and the soft moonlight

A P O L L O

8

PROLOGUE

August 1968

THE LAST THING FRANK BORMAN NEEDED WAS A PHONE CALL WHEN he was trying to fly his spacecraft. No astronaut ever wanted to hear a ringing phone when he was in the middle of a flight, but when the space-craft was an Apollo, any interruption was pretty much unforgivable. The Apollo was a beautiful machine—so much bigger, so much sleeker than the corrugated Mercury and Gemini pods that all of the other Americans who had ever been in space had flown. But the Mercurys and the Geminis had a perfect record: sixteen launches, sixteen splash-downs, and not a crewman lost. The Apollo, on the other hand, was already a killer: only eighteen months ago, three very good men had died in the first ship, before it even got off the launchpad.

So when Borman was trying to fly the cursed machine, he needed to pay complete attention. And now, at precisely the wrong moment, there was a call for him.

In fairness, Borman was not actually in midflight when the phone rang. No one had yet taken an Apollo into space; that would be folly until the ship was proven fit to fly, which it most assuredly had not been. For now, he was merely on the factory floor at the North American Aviation plant in Downey, California, where all of the new Apollos were being built. But Borman *was* sitting in the cockpit of an actual Apollo ship, one that was currently known as "Spacecraft 104," though it would soon enough be called Apollo 9. And it *was* his ship, the one he would command if it ever got off the ground. If it did fly, Borman's place would be in the left-hand seat—the ranking seat—and that suited him just fine. His crewmates, Jim Lovell and Bill Anders—exceptional men, both— would be in the center and right seats. Lovell and Anders were with him today, in fact, and the work they were doing was every bit as difficult as his own.

Apollo 9 was scheduled to launch in approximately nine months, which put Borman and his crew in the stretch run of their training. That schedule, however, depended on Apollos 7 and 8, the first two manned flights of the Apollo series; both had to get off the ground and bring their crews home whole and well. All three of the flights would be limited to Earth orbit, and to Borman's way of thinking, that was a shame. It was the boiling summer of 1968, and the world had spent much of the year bleeding from countless wounds: multiple wars, serial assassinations, riots and unrest from Washington to Prague to Paris to Southeast Asia. The Soviets and the Americans, again and always, were staring each other down in hot spots around the globe, while American boys died in Vietnam at a rate of more than a thousand each month.

A flight to the moon—which President Kennedy had once promised would happen by 1970—would have been a fine and bracing achievement right about now. But Kennedy was five years dead and three Apollo astronauts were eighteen months dead and the entire lunar project was flailing at best, failing at worst. Most people believed that if American astronauts reached the moon at all, they wouldn't get there for years.

Still, Borman had his mission, and he and his crew had their ship. And today they were inside it, running their flight drills and doing their best to get the feel of the machine. All of the Apollos looked the same and were laid out the same way, but spacecraft were like aircraft. Pilots could feel their differences—in the give of a seat or the grind of a dial or the stickiness of a switch that had a bit more resistance than it should. Each spacecraft was as particular to each astronaut who flew it as a favorite mitt is to a catcher, and you had best know your ship well before you took it into space.

Now, as Borman, Lovell, and Anders lay in their assigned seats in their small cockpit, working to achieve that flier's familiarity, a technician popped his head through the hatch.

"Colonel, there's a phone call for you," he said to Borman.

"Can you take a message?" Borman asked, annoyed at the interruption.

"No, sir. It's Mr. Slayton. He says he has to talk to you."

Borman groaned. Slayton was Deke Slayton, the head of the astronaut office and the man who made the crew selections and assigned all of the men to their flights. That power came with the understanding that he could always *un*-assign you to a flight if he chose. When Slayton rang, you took the call.

Borman crawled out of the spacecraft and trotted to the phone. "What is it, Deke?" he asked.

"I've got something important I need to talk to you about, Frank."

"So talk. I'm really busy here."

"Not on the phone. I want you back in Houston now."

"Deke," Borman protested, "I'm right in the middle of—"

"I don't care what you're in the middle of. Be in Houston. Today."

Borman hung up, hurried back to the spacecraft, and told Lovell and Anders about the call, offering only a "who knows" shrug when they asked him what it meant. Then he hopped into his T-38 jet and flew alone back to Texas as ordered.

Just a few hours after he was rousted from his spacecraft, Borman was sitting in Slayton's office. Chris Kraft, Borman noticed with interest, was there as well. Kraft was NASA's director of flight operations; as such, he was Slayton's boss and Borman's boss and almost everyone else's boss, save the top administrators themselves. But today he remained silent and let the chief astronaut talk.

"Frank, we want to change your flight," Slayton said simply.

"All right, Deke . . ." Borman answered.

Before Borman could say anything else, Slayton held up his hand. "There's more," he said. "We want to bump you and your crew from

Apollo 9 up to Apollo 8. You'll take that spacecraft since it's further along—and you'll fly it to the moon."

Then, as if to make clear that the astounding statement Borman had just heard was really what Slayton meant to say, he put it another way: "We are changing your flight from an Earth orbital mission to a lunar orbital." He added, "The best launch window is December twenty-first. That gives you about sixteen weeks to get ready. Do you want the flight?"

Borman said nothing at first, taking in the sheer brass of what Slayton was proposing.

Before Borman could fully gather his thoughts, Kraft spoke up: "It's your call, Frank."

That, all three men knew, was entirely true—and entirely untrue, too. Borman was a soldier, a West Point graduate and an Air Force fighter pilot. He had never had an opportunity to fight in a hot war, but the space program was a race with the Soviet Union and a critical part of the Cold War. A battlefield assignment—no matter what kind of battlefield— was not something he could possibly turn down.

The way Borman saw it, circumstances might warrant your saying no to a dangerous assignment, and your commanding officer might forgive you for saying no, but if you hadn't signed up to fight, then why did you become a soldier in the first place? And if you hadn't joined the space program to fly to the moon when your boss and your nation and— somewhere in that long chain of command—your president were asking you to, well, maybe you should have chosen a different line of work. The Apollo spacecraft might not be up to the job, the flight planners who had the same sixteen weeks to get ready for a mission to the moon might not know exactly what they were doing, and in the end three more Apollo astronauts might wind up dead. But death was always a part of the piloting calculus, and this time would be no different.

"Yes, Deke," Borman said. "I'll take the flight."

"And Lovell and Anders?" Slayton asked.

"They'll take it, too," Borman responded briskly.

"You're sure about them?"

"I'm sure," Borman answered. Then he smiled inwardly. He could only imagine the look on Lovell's and Anders's faces if he had flown back

to Downey and told them that they had all been offered the chance to go to the moon before Christmastime and he had answered, *No thanks*.

☆ ☆ ☆

There was no established way for a man to tell his wife he was going to the moon. A man could tell his wife he was going to sea or going to war; men had been doing that for millennia. But the moon? It was a whole new conversation.

Reporters loved stories about astronauts coming home with the exciting news that they had been assigned to a mission. They especially loved it when the astronauts would let them photograph the family—the brave man posing with a world atlas in his lap, tracing the path his spacecraft would make around the Earth, the children sitting on either side of him, the wife standing over his shoulder beaming down at the entire tableau. Surely a map of the moon would make an even grander picture.

There would be no moon maps in Frank Borman's house, however. When the commander got word that his assignment had been changed from an Apollo 9 Earth orbit to an Apollo 8 lunar orbit, he came home and told his wife, Susan, the news; she looked at him and said, "Okay." He then told his boys, who were seventeen and fifteen; they looked at him and said, "Okay," too. That was the way it had gone the first time Borman had flown into space, three years earlier, and that was the way it had always gone when Borman was given a dangerous assignment. With or without the pretty pictures of the families with their atlases, it had gone that way in all of the other Houston households, too, when, sixteen times before, American astronauts had prepared for a flight into space.

But this time was different, because none of those other astronauts had flown to the moon, and a moon mission was exactly what Susan's husband had just accepted. Still, she and Frank would find a way to manage this challenge, just as they had managed every difficult challenge they'd faced in eighteen years of marriage. They even had a shorthand for their preferred approach, and it went like this: "The custard is in the oven at 350 degrees."

Frank would mention the custard when he was requesting combat

duty or trying out a dangerous new plane; Susan would mention the custard when they were looking at a new house or sending the boys to a new school. They both took comfort in the little incantation. What the custard meant—with its sense of domestic coziness—was that Susan would tend to the home and Frank would tend to the flying. And if they both did that, while taking care not to overstep to the other person's patch, the custard would come out just fine.

But this new challenge, Susan realized, might require something more than the custard. She had been an astronaut's wife for a long time, and she realized at a primal level that something about this mission was particularly troubling, something that went beyond the obvious risks most people were considering. Frank wouldn't be landing on the moon; in fact, the spidery lunar module that would be needed for that final step in the lunar project hadn't even been fully built yet. But an orbital mission posed its own dark peril.

The most important piece of hardware on an Apollo spacecraft was its engine—the big, roaring blunderbuss at the rear of the ship, the machine that some of the more superstitious people in NASA referred to merely as The Engine, the same way they would say The Queen or The President or The Moon itself, with the "The" conferring an almost supernatural specialness. Borman, Lovell, and Anders referred to the engine by its brisk, clinical initials: to them it was the SPS, short for "Service Propulsion System." The label nicely expressed its job, which was to propel and to serve.

A machine that played so central a role in a lunar mission could not possibly fail. But what if it did? That was the worry. Once the rocket that gets you off the ground and hurls you toward the moon is done with its job and falls away, the main means of propulsion you have left is the SPS. If you were planning to orbit the moon, you would need it to fire at least twice: the first time to slow you down, so you'd surrender to the moon's gravity and become a lunar satellite, and the second time at the end of your visit, so you'd speed back up and peel off for Earth. If the engine failed the first time, the mission would be wrecked but the crew might survive, whipping around the moon and coming home. If it failed the second time, the crew would be trapped in lunar orbit. They would circle the moon as it circled the Earth, on and on, forever and

ever. Sealed inside a metal sarcophagus, the astronauts would never come home, but they'd never descend for a conclusive crash into the lunar surface, either. That engine failure would effectively ruin the moon: no one would ever be able to look up at it again and not be aware of the three dead men.

So Susan decided to talk to NASA's director of flight operations. Apart from her husband, Chris Kraft was perhaps the only person in the space program with whom she could be herself. He had a sandpaper temperament and a pitiless honesty that Susan found refreshing. And since the NASA families living in the communities around the space center in Houston were a sociable bunch, she figured she'd have a chance to talk with him before long. Sure enough, shortly after Borman got his lunar assignment, Kraft dropped by their house one evening, and Susan seized a private moment.

"Chris, I'd really appreciate it if you'd level with me," she said. "I really want to know what you think their chances are of coming home." It was a straight-up question and she held his eyes, insisting that he give her an equally straight-up answer.

Kraft studied her face. "You really mean that, don't you?" he asked.

"Yes," Susan said, "and you know I do."

Kraft did know. "Okay," he said directly. "How's fifty-fifty?"

Susan nodded. She had suspected as much.

ONE

Mid-1961

FRANK BORMAN WASN'T FEARLESS—NO GOOD PILOT WAS. BUT MOST of the fear he had started out with was shaken out of him on the day in 1961 when he nearly died in the skies over Edwards Air Force Base, in southern California. Death while flying was not Borman's plan, any more than it was the plan of the pilots who did die that way in that place at around that time. But in the business of test-flying jets, it was inevitable that every now and then one of them would fly straight into disaster, tumble out of the sky, and, to use the no-big-deal phrasing preferred by test pilots, make a hole in the ground.

Borman knew the odds as well as anyone, but he had been angling for Edwards for a long time, ever since his West Point graduation in 1950. Like most of the men flying here, he had not traveled a direct route to the skillet of the California desert. He had jumped from post to post— Nevada, Georgia, Ohio, the Philippines—all with his young wife, Susan, by his side. She had agreed to marry him straight out of the military academy and had known in advance the peripatetic life of a young Air Force pilot, though she hadn't fully understood that it would be quite *that* peripatetic.

Still, Edwards was what Borman wanted, and so Edwards was what Susan wanted. When he had applied for the transfer from the Philippines to California in 1960, he'd known he would be up against ferocious competition, with no guarantee that he'd even make it to the top tier of applicants, much less become one of the fliers selected. When he did get the transfer, he was thrilled at the news, though it wouldn't serve him well to look thrilled when he arrived in California. The man who made the Edwards flight assignments was Chuck Yeager, the thirty-seven-year-old lieutenant colonel and World War II fighter pilot who had long caused military fliers to feel equal measures of inspiration and terror.

Yeager, the first man to break the sound barrier, in a Bell X-1 rocket plane in 1947, had been famous even before achieving that long-elusive feat. Shot down over France in 1944 during his eighth combat mission, he was taken prisoner and placed in a POW camp. He escaped just two months later, quickly rejoined his squadron in England, and, six months after that, scored the fighter pilot's coveted ace-in-a-day distinction, shooting down five enemy planes in a single mission. There wasn't much the young hot dogs he had brought to Edwards could do to impress him.

"There's no such thing as a natural-born pilot," he would tell the men who believed they were exactly that. "If you can walk away from a landing, it's a good landing. If you can use the airplane the next day, it's an outstanding landing." Keep it simple, Yeager told them, and you may stay alive.

Borman reckoned he could do more than that. In the years before getting to Edwards, he had flown the F-89 and the T-33 and the T-6 and the F-84 and even the F-104. He had flown just about every plane he could get his hands on, and not one of them had been too much for him.

One morning soon after getting to Edwards, Borman hopped into an F-104, intent on trying out a maneuver of his own invention. He called it a zoom flight, a nod to the extreme altitude and trajectory the exercise would involve. Both the plane and the new maneuver were their own kinds of trouble. Together, they were a two-headed beast.

The F-104 was a relatively new plane; test-flown for the first time just five years earlier, it had been certified fit to fly only three years back. The jet was designed to be extremely lightweight and to fly at extremely high speeds. Just fifty-five feet long, it had a wingspan of less than twenty-one feet and an airframe made of such light aluminum that, for practical purposes, the whole assembly was little more than an engine with a chair on the front. The F-104's wings were positioned so far back on the body that the man in the cockpit could not see them without a rearview mirror. And the leading edges of the wings were so fine—just .016 of an inch thick, or about the same as a razor blade—that ground crews took to covering them with protective strips lest they brush past one and cut themselves.

All of these innovations made it possible for the plane to achieve a reliable speed of Mach 2.2—or more than twice the speed of sound— with the right altitude and wind conditions. The aircraft's design, however, did not deliver much in the way of maneuverability. Supersonic speed means a wide turning radius, and a streaking jet is a little like an ocean liner that swings hard to port but doesn't actually complete its turn until miles later. One pilot who had flown the F-104 and not cared for its handling described steering the plane as "banking with intent to turn." That was funny enough in the skies over Edwards Air Force Base, but the humor would be lost in a dogfight with a Soviet MiG. You might win the fight if it was merely a chase in the flat, though it would turn ugly once the MiG began to weave.

There were no MiGs over California, however, and Borman's zoom flight didn't call for any weaving. Instead, it would begin as an ordinary climb to 40,000 feet, nearly eight miles above the desert floor. At that point, he would light the afterburner, which would slam him back into his seat and punch the plane up to 90,000 feet.

That kind of altitude presents very particular challenges. Once you get there, neither the afterburner nor the engine work anymore, both of them strangled by the thin atmosphere. Even if you could light your engine, it would be a very risky move. The F-104's engine was air-cooled, and with little air available, an engine that was lit up might also blow up.

So now you are seventeen miles above the desert with no working form of propulsion. Climb any higher and those stubby wings on your plane are going to become useless too, since the few wisps of effective atmosphere that remained at 90,000 feet would be gone. Instead, you arc over, plunge back down, relight your engine at 70,000 feet, and come on in.

The zoom flight was a fine way to test the mettle of both the plane and the pilot. And if you had the guts to try it, it was an awful lot of fun.

Borman had flown the zoom several times before, always without incident. But this morning would be different. No sooner had he reached 40,000 feet on his upward climb than the engine did what engines are never supposed to do: it exploded. The bang was audible, the jolt was considerable, and the red fire light on the instrument panel confirmed that both the aircraft and the pilot were in mortal trouble.

What the book called for at a moment like this was cutting the engine—or whatever was left of it—to reduce the risk of fire. What the book definitely did not call for was ejecting. Yes, staying in the plane might be deadly, but hitting the air when ejecting at supersonic speeds would be akin to hitting the side of a building. Dead-sticking, or gliding without power down to the runway, was a theoretical possibility, but dead-sticking from 40,000 feet while traveling at Mach velocity was not even worth attempting.

With no even remotely good options, Borman was left to choose the least awful course of action, which was to try to restart the engine. The plane's half-wrecked piece of machinery might or might not have the wherewithal to function as it should; it most surely, however, had the power to explode again, pulling the plane apart entirely this time.

His decision made, Borman hit the ignition. The engine banged but burned. The fire light came back on, and the plane clanged and shook. Still, the engine ran for three full minutes, enough to bring the dry lake bed into view. At this point, Borman could have ejected—indeed, he knew he *should* have ejected—but he had brought the plane this far, and by now it was a matter of principle to land it on the runway in one sorry, faithless piece. Maybe the engineers could take it apart and see how and why it had gone wrong.

Borman did land the smoldering aircraft, and then, as soon as he quit taxiing, he hopped from the cockpit and abided by one more iron rule of piloting: he ran as fast and as far as he could from the plane. The fire trucks, which were already converging, would deal with the rest.

Once he was safely out of harm's way and in the base office, Borman picked up the phone and, fulfilling his part of the flier's marital bargain, called his wife. Susan was well aware of how risky her husband's work was; recently, in fact, she had witnessed a collision over Edwards between an F-89 and a T-33. She knew that Frank flew both planes, so she had dashed to the scene of the wreck to see if he was one of the pilots. She was stopped well before she got there by a military sentry, who told her she had to go home. She went instead to the home of a neighbor, a woman married to an officer and someone who had been through this terror before. The neighbor took her in and calmed her down and explained the Edwards rules to her. "That isn't how it's done," she said. "You have to sit and wait."

Frank had not been in either plane that time, but absorbing the lesson of that morning had not been easy for Susan. So he had promised her that in the future, whenever an accident happened, he would contact her as fast as he possibly could, and he did so today.

"You're going to hear about a problem, and you might have seen the fire trucks go out," he told her as soon as she answered the phone. "It was me. And I'm fine."

Susan, now practiced in the protocols, said that she was relieved to hear it, that she always trusted that he would bring his planes in safely, and that she was very glad to know he was all right. What she didn't do was give voice to the words that usually came into the minds of test pilots' wives in such moments: "This time."

☆ ☆ ☆

Borman's abiding love of flight began in childhood, but his career in the air almost came to an early end. And it was his well-loved Air Force that tried to ground him.

Born in 1928, the only child of Edwin and Marjorie Borman, Frank was supposed to grow up in Gary, Indiana, where his father ran a successful garage business, a point of pride in the early years of the Great Depression, when so many other families were struggling. But the trick to living happily in Gary was overlooking the persistent chill and leaden dampness that often settled on the area. Though most Indianans could tolerate the weather, young Frank couldn't. Sinus problems, repeated colds, and mastoid infections kept him out of school so regularly that the family's doctor warned his parents that if the boy didn't get somewhere dry and warm fast, he might grow up with no hearing at all.

So the family abandoned their life in Gary and moved to Tucson, Arizona. Frank promptly got well, thrived in school, and took to building and flying model airplanes, then more model airplanes, and then many more still. When he was a senior in high school, the year after the end of World War II—a conflict in which airpower played a major role—he decided that flying real airplanes was what he wanted to do with his life. The best route to the best planes, he knew, was to get himself accepted to West Point, and from there to make his way to the Air Force.

By the time Borman decided to apply to the military academy, however, it was too late in the year to submit his application and get the necessary recommendation to West Point from his congressman. While trying to decide what to do next, he was approached by a local judge who knew him from the neighborhood. The judge had a son younger than Borman who was turning out to be something of a wild child, always on the brink of getting into trouble with the law. Impressed by Borman's straight-arrow ways, the judge wondered if he might teach his son a wholesome hobby like model-airplane building, and get him interested enough to keep him off the streets.

"I'll try, sir," Borman said simply.

"That's all I can ask," the judge responded. "It'll be to your credit if you succeed, but no one will blame you if you don't."

As it turned out, he did succeed—love of flight being a pretty easy disease to catch in 1946, and Borman being a very infectious carrier. He and the judge's son spent most of their after-school and weekend hours working together on their planes, which slowly brought the younger boy to heel.

The judge, who knew of Borman's West Point ambitions, offered to show his appreciation by contacting the local congressman himself—a veteran politician named Dick Harless—and asking him to write the needed letter of recommendation. Harless agreed, but no congressman could change the fact that it was now very late in the West Point application process, and the best the military academy could offer Borman was a third-alternate slot: three boys who had already been accepted would have to decline the academy's offer. Improbably, the long shot came in, as, one after another, the potential cadets decided that maybe they weren't West Point material after all. Borman, to his own amazement, took his place as a shaven-headed plebe in the class of 1950.

As he'd suspected he would, Borman loved every single thing about West Point. His family's survival during the Depression had depended on his father's inexhaustible dedication to his work; Borman had absorbed that lesson, applied it in high school, and now did so again at the academy. He loved West Point's head-cracking academics and ferocious discipline and the deep camaraderie that came from standing on the lowest rung of an exceedingly hierarchical system.

Unlike a lot of the other plebes, Borman even learned to appreciate the self-control that came from tolerating the hazing at the hands of more senior cadets, though that part was not easy. There was the business of eating in silence in a full-brace, straight-backed, eyes-ahead position, and complying uncomplainingly when an upperclassman would unmake a perfectly made bed and have him remake it. And he learned, too, a subversive secret about West Point: occasionally there were moments—well considered, precisely chosen—when it paid greater dividends *not* to comply.

Early in his first year, as Borman stood in ranks, an upperclassman known as a relentless bully was inspecting the plebes, paying particular attention to their shined shoes. When he reached Borman, he stopped. "Mr. Borman, those shoes don't look good," he declared.

"Yes, sir," Borman answered, staring straight ahead and, as he'd been trained, avoiding eye contact.

"But now they're going to look worse."

With that, he lifted his foot and slammed the heel of his own shoe on top of Borman's.

Borman remained standing, ignoring the blaze of pain that ran through the fine bones on the top of his foot. Then he hissed quietly, "You son of a bitch. If you ever do that again, I'll kill you."

Silence came from the upperclassman, as well as from the plebes nearby, who had heard what Borman said but dared not acknowledge it. Such insubordination could wreck a West Point career before it even got started, and everyone knew it. The upperclassman stood and glared—and then moved on. Borman's defiant bet paid off; as best he was ever able to piece it together, his tormentor already had a reputation for abusive hazing and could no more afford a reported incident than could Borman himself.

Borman's rise at the military academy was swift in every respect save one: sports, specifically football, which at West Point was pretty much the only sport that counted. He might have had the grit for the game, but at just five foot seven and 155 pounds, he definitely did not have the size. Bantamweight was good for a pilot, not a lineman. Nevertheless, Borman joined the team anyway, in the only capacity in which they would have him—as manager, which effectively meant equipment boy, albeit with some organizational and scheduling responsibilities.

All the same, he thrived in the job. Partly that was because he got to work with the likes of head coach Earl Blaik—equal parts man and monument in college football—and a young offensive-line coach named Vince Lombardi. Both were tactically brilliant, though Blaik maintained a rigorous cool while Lombardi was a man of raw emotion. Borman saw himself in Blaik, but he also admired the happy madness that was Lombardi. Most of all, however, he liked the simple yet vital work the job of manager required him to do. Although he could not contribute on the field, he could bring a necessary order to the entire enterprise, without which everything else would come to ruin.

Ultimately, the young cadet became so assimilated to military life that when his four years at West Point were done, he graduated eighth in his class of 670 students, an accomplishment impressive enough to earn him his longed-for assignment to the Air Force. Upon graduation, he received orders to report to Nellis Air Force Base, in southern Nevada, where he would train to fly the F-80 fighter jet, before shipping out sometime in 1951 to combat in Korea.

Borman had achieved precisely the trajectory he had planned for himself, with one critical exception. Unlike many of his classmates, he had not had a serious girlfriend for most of the time he'd been at West Point; thus he had no wife now that he'd graduated. And for that he had no one to blame but himself.

Back in high school, Borman had fallen in love with a blond-haired girl named Susan Bugby. At first he had been too nervous to approach her, but once he did, she returned his interest, and before long he had little doubt that he'd met the girl he wanted to marry. He was new to romance, though not to good sense: if you loved a girl and she loved you, why would you go looking for someone else? Before Borman left for West Point, he and Susan informally agreed that they would marry upon his graduation. In his first year at the military academy, however, Borman got full of himself—or at least full of the challenging academics and the monastic life that seemed fitting for a cadet—and he broke off their relationship. The way he saw it, a cadet had time for West Point or for love, but not both.

Almost immediately, he regretted his decision, and he missed Susan keenly throughout the rest of his time at the military academy. Shortly before he graduated, he wrote to her at the University of Pennsylvania, where she had gone to study dental hygiene. In his letter, he asked whether they might set things right now that his time at West Point was ending. She wrote back, saying that while she still cared for the soon-to-be lieutenant, her circumstances had changed. She was involved with a boy back home, someone who did not seem inclined to put her aside if his career became too interesting.

Borman was crestfallen, but thanks to a combination of competiveness, stubbornness, and at least a little youthful conceit, he didn't accept her answer as final. He continued to write Susan letters in the months that followed, and to his delight and surprise, she continued to write back. He asked after Susan's family and her education; she asked after his parents and his West Point experiences. He told her—just in case she was wondering, though of course he understood that she wasn't—that he had dated no one else seriously in all the time he'd been away. She wrote back to say that that was fine, but reminded him that she couldn't say the same.

And then, after a time, she wrote to say that in case Frank was interested, her circumstances had changed again and it seemed that she didn't care for the other boy as much as she'd thought and they had parted. Borman wrote back immediately to say that he was very, very happy to hear that. He also told her that he would be coming home to Tucson for a visit before he had to ship out to Nellis, and suggested that perhaps they might get together for dinner. Susan agreed.

Borman chose a quiet but elegant place in the desert about fifteen

miles outside of the city. It was a hideaway with an Italian restaurant and
a swimming pool where married—and, no doubt, unmarried—couples
could go to be alone. There was also, Borman knew, dance music.

Over the dinner and during the dancing he talked about his plans
for the future and his hopes for a family—and how none of that would
be complete with anyone else but her. He told her how foolish he had
been to toss aside what they had shared. He swore that if he had it to do
a million times over he would never, ever make the same mistake
again. She seemed moved and admitted that she felt the same about him
as she always had. Borman breathed easier. He felt certain that now there
was only one thing left to do, which was to ask Susan to marry him—
although he wasn't certain just how and he wasn't certain just when.

They left the restaurant together and began the drive back to Tucson.
At some point along the way, Borman decided that *this* was the proper
moment and that this otherwise unremarkable spot on the road somewhere between the desert and the city was the proper place. So he pulled
over and said, in as straightforward a way as he could muster, "Let's get
married."

With equal directness, Susan responded with a single word: "Wonderful."

Frank beamed, reached into his jacket pocket, and produced the
ring he had bought recently against the chance that tonight might be his
opportunity to undo the terrible mistake he had made a few years before.
She accepted it.

And so—twenty-two, love-drunk, and cussedly stubborn—Lieutenant
Borman found himself engaged.

☆ ☆ ☆

Nevada's Nellis Air Force Base was exactly what Borman expected it to
be: a passable enough place for a young military officer and his new
bride to make a home, and an extraordinary place to learn combat flying. And it was clear to all of the fliers on the base that their skills would
be needed soon.

Just that June, at about the same time Borman and his classmates
were graduating, the Soviet-backed North Korean army had launched a
massive offensive across the 38th parallel, the ostensible no-go zone that

separated the two halves of the Korean Peninsula. Within three days, the 75,000-man onslaught swept aside the American-backed South Korean defenses and arrived in Seoul.

The men at Nellis were anxious to get into that fight, and their training was intensified so that they would be battle-ready in as short a time as possible. Borman, whose star had shone so brightly at West Point, was determined to make his mark as an aviator as well. And that led him to do a very dumb thing.

One afternoon, several months after arriving at Nellis, Borman hopped into an F-80 to run some dive-bombing drills, just to sharpen the skills he would need once he got into combat. It was a fine way to spend a free hour or two, except on a day when a pilot had a bad head cold. Borman's ears and sinuses had given him little trouble since his family quit Gary so long ago, and if he had suffered any lasting damage from his early years of chronic infections, it hadn't shown itself. So Borman—with head congestion familiar enough to him after his afflicted childhood that he didn't even notice it—went off flying.

At the top of a climb, just as Borman was beginning a dive, his head exploded, or felt as if it did. A lightning bolt of pain erupted from somewhere deep between his ears; he managed to defy the natural response to pain, which is to grab the thing that hurts—grabbing your head when you're flying an F-80 at 600 miles per hour is simply not an option.

Instead Borman gritted his teeth, nursed the plane and himself down to the ground, and immediately went to see the doctor on the base. The pain now seemed to be localized in one ear; peering inside that ear, the doctor made the lethal little tsking noise doctors make when they suspect that something is seriously wrong and then discover that the situation is even worse than they'd feared.

"The eardrum," he explained to the young lieutenant, "is made of three separate layers. You've ruptured them all."

Whatever Borman had been thinking when he'd chosen to fly on a day when his head was badly clogged was his own affair, the doctor said, but the damage he had done was serious. The ear might or might not heal; either way, the doctor couldn't offer a meaningful prognosis until Borman came back for a follow-up visit in about six weeks. In the meantime, his file would be stamped DNIF, or "duty not involving

flying." For a pilot on the rise, those four letters were as good as an epi-
taph.

Borman protested, but the doctor held fast. The young aviator
explained that his unit would be shipping out for Korea in less than six
weeks; that may well be true, the doctor said, but if so, his fellow fliers
would be going without him. Ultimately, Borman was indeed left behind,
and when he returned to see the doctor six weeks later, the three ruptured layers were still blown.

Not long after, Lieutenant Borman—certified jet pilot, graduate of
West Point—did ship out for the Pacific, but it was to a peaceable billet
in the Philippines. The post was so peaceable, in fact, that he was able
to bring along his wife, Susan, and their infant son, Frederick, born just
weeks before. They would all be assigned a spot in base housing just
right for a young family. His new assignment would be director of roads
and grounds; effectively, he would be the base's chief of maintenance.
There might be a more humiliating title for a grounded flier than one that
actually included the word "grounds," but Borman did not care to try to
come up with one.

☆ ☆ ☆

Life in the Philippines was not merely as bad as Borman had feared it
would be, it was vastly worse. His hangar had been replaced by a garage;
his F-80 had been replaced by road graders and steamrollers. After
allowing a decent interval to go by, he applied for a return to flight status, eardrum notwithstanding. His application was rejected. He applied
for a transfer to Korea, so that he could at least work as a forward air
traffic controller. That was bounced back at him, too. He applied to quit
the Air Force altogether and transfer back to the regular Army; at least
he would still have his West Point diploma and his lieutenant's commission. Again the answer was no.

Now desperate, Borman visited a doctor on his base in the Philippines; he, too, looked in the ear and made the same sympathetic clucking
noise every other doctor had made over the last year. This doctor, however, also told him about a woman he'd heard of in Manila, a specialist
in otolaryngology, who had developed a technique for fixing ruptures in
the ear. She would deposit pellets of radium in the eustachian tube, and

the radium, through a mechanism she could not explain, seemed to speed healing. Borman took the bus to Manila, submitted to the procedure, and waited the prescribed time before returning to the base doctor, who told him that one layer of his eardrum had healed and even developed a thick layer of protective tissue. But that still left two layers wrecked. And that, in turn, left the DNIF in place.

Finally Borman appealed directly to the squadron commander, Major Charles McGee, a Tuskegee airman who had flown and fought extensively in World War II. Being an airman was hard; being a black man in midcentury America was, Borman assumed, at least as hard. A man who had met both challenges with as much aplomb as McGee was someone to be respected—and someone Borman assumed would play straight with him. He approached McGee and explained that he was absolutely certain that his one-layer eardrum was up to the job of flying, at all altitudes, in all conditions, with the cockpit either pressurized or unpressurized. The problem was, the Air Force was refusing to give him the chance to go up and find out for sure. He understood that if he flew and the eardrum blew again, he would have to accept a permanent grounding. But if it didn't blow, he could be back in the sky.

McGee agreed that it was worth finding out, so he took Borman flying. First they went up in a T-6, then in a higher-altitude T-33. McGee put both his passenger and his passenger's battered ear through as challenging a ride as a pilot could bear. Borman tolerated the changing pressures and the dizzying swoops with no difficulty and no pain. When they landed, McGee smiled at Borman and told him, "Better go see the doctor again." Before Borman could dart off and do just that, McGee added a warning: "And tell him the truth."

The doctor inspected Borman's ear and offered the usual unremarkable diagnosis: the one healed layer of the eardrum was still intact, but otherwise there had been no improvement. Borman cut him off before he could finish.

"Doc, you might as well know I've been flying with McGee," he said. The doctor looked both surprised and displeased, and Borman came out with the rest of it: "Even in a T-33," he said, with no small bit of pride.

"I'll need to get McGee to confirm that—especially that T-33 ride," the doctor said.

He then dismissed Borman and called McGee, who told him that yes, everything the eager young lieutenant had said was true. The doctor was skeptical—*he* was the medical man, after all. But McGee was a flier, and he had spent a good part of a grueling day with another flier, and a fine one. Perhaps, the squadron commander suggested, a man who so badly wanted to serve his country and had now proven his fitness to do so ought to have his wings returned to him.

The doctor evidently agreed. Shortly afterward, the official notification arrived in the Borman home. "In the case of First Lieutenant Frank Borman," the document read, "subject has been returned to flying status."

☆ ☆ ☆

In the decade that followed, Borman pursued his love of flight with a near-consuming fever. He and Susan and their now two young sons—Edwin arrived just nineteen months after Frederick—moved from base to base, hopscotching the United States as the needs of the military demanded. By the time he was practicing zoom flights at Edwards Air Force Base in 1961, as enviable a billet as a pilot could hope for, he had at last become the highly accomplished flier he had long wanted to be. One piece, however, was still missing.

Borman had been too young to fight in World War II and had been grounded before he ever got a chance to get to Korea; now, as a thirty-three-year-old pilot with a wife and family, he knew that his chances of engaging in the combat he'd been training for his entire adult life were rapidly diminishing. A shooting war with the Soviets, which had barely been averted in 1956 when Moscow sent tanks to crush a short-lived democracy uprising in Hungary, still seemed to be a real possibility. Yet if the call to arms came, it would likely not go out to him. There were younger fliers who would surely be sent first. Borman had missed his window.

But if there were hot warriors, as there had been throughout all of humanity's bloody history, there were now Cold Warriors, too. This was a whole new kind of fighting. The Cold Warriors drew up the plans and trained the soldiers who might have to fight if the Soviets tried a stunt like Hungary again. They manned the Atlas and Titan and Minuteman missile silos all over the American West, ready to respond if Soviet missiles started flying. They flew the Strategic Air Command's B-52s and sailed the ballistic missile submarines, keeping America's nuclear assets mobile twenty-four hours a day.

And then there were the seven Cold Warriors everyone knew best, the ones who had been hand-selected from the military in 1959, dressed in silver pressure suits, and taught to fly not jets but rockets. These were the soldiers sent off to beat the Soviets in the highest, fastest combat of all: the competition for space. They were the first fighters in a different kind of war, and if you thought that flying rockets didn't count as real combat—that climbing on top of a ninety-five-foot Atlas booster full of explosive fuel was not at least as big a risk as flying into battle—well, you didn't know much about calculating odds.

The country's first astronauts had become famous and, by pilots' standards, rich, with the magazine contracts and the free Corvettes people kept giving them. But the individual glory accorded to these warriors didn't mean a lot to Borman. The collective glory of the Air Force did, however, and in 1961 the branch of the military he loved best was underperforming. Of the first seven astronauts, only three, Gus Grissom, Deke Slayton, and Gordon Cooper, were Air Force. That might be good enough for the Navy or the Marines, but the Air Force was the only branch of the service that had flying in its very name. In Borman's view, three out of seven simply did not cut it. Now, even before all of the original astronauts had flown, the call was going out for a second class of recruits. This time nine of them would be selected, and they would be flying not the little one-man Mercury spacecraft but the two-man Gemini and later the three-man Apollo—and the Apollos were the ships that would go to the moon.

At Edwards Air Force Base in California and Wright-Patterson Air Force Base in Ohio and Naval Air Station in Patuxent River, Maryland, pilots were quietly submitting their applications to NASA, the National Aeronautics and Space Administration. Borman did the same. The other branches of the military sent mixed signals to their men about whether it was right—or even loyal—to walk out on their current assignments, which was why so many of the applications were submitted so discreetly. The Air Force, by contrast, aggressively encouraged its men to volunteer.

In case there was any doubt about that, General Curtis LeMay, the Air Force chief of staff himself—a concrete bollard of a man who had flown bombing missions in both Europe and the Pacific during World War II—summoned the officers who had submitted their applications to NASA for a sit-down in Washington. Borman was among them.

"I'm hearing that some of you think you'll be deserting the Air Force," he said in that signature growl that could seem unfriendly, mostly because it was. "You're not deserting the Air Force, and you're not ducking combat. The Cold War is real, real as any war. Go fight it—and make the Air Force look good."

That was all the invitation and forgiveness Borman needed. Not long after his application went out, he was called in for weeks of grueling physicals and other trials demanded by the space program. He had no fear of passing the flight tests and skills tests and intelligence tests, but the medical tests were another matter. While his single-layer eardrum might have been good enough for jets, he had no idea whether it would be deemed suitable for space. He lived in dread of the moment when the first NASA doctor would examine his ears.

The day for his medical examination arrived, and as soon as one of the NASA doctors stuck a scope inside the damaged ear, he emitted a low whistle of disbelief.

"Get a look at this," he called to another doctor, who came over, took the scope, and made the whistling noise, too. This little scene was repeated two or three times, until at last the man who appeared to be the head doctor looked through the scope.

"Young man," he asked Borman, "does that ear bother you?"

"No sir it doesn't bother me it doesn't bother me at all," Borman answered with no audible pause between any of his words.

"Well," the head doctor replied after a thoughtful moment, "if it doesn't bother you, it doesn't bother me."

And that, to Borman's utter amazement, was that.

Borman went back to Edwards, where he could do nothing but wait for NASA's decision. He did not have to wait long. One morning in the spring of 1962, he was on-base and received a phone call from Slayton, who was now the head of the astronaut office, and learned that the decision had been made: yes, he was going to become an astronaut. He hung up the phone, pumped his fist in triumph, and drove straight home to Susan. The moment he walked through the door, she could tell from his face that something very good had happened. She inclined her head in that "out with it" way of a wife who knows her husband well.

"Well, look," Borman said, suddenly feeling more modest about his news than he'd expected to. "I was selected."

Susan did not have to ask for any more information. She threw her arms around him and hugged him tightly. It was one more reassignment in the life of an Air Force officer's family—but an assignment like no other.

Borman knew that the next thing—the decidedly more awkward thing—he had to do was go tell Chuck Yeager. He marched over to Yeager's office and waited until the great man called him in.

"Colonel, I just got some good news," Borman said.

"What's that?" Yeager said, looking up from his desk with little interest.

"I was just selected to go to NASA and join the astronaut corps."

Yeager nodded; for a moment or two he was silent. "Well, Borman," he said finally, "you can kiss your Air Force career good-bye."

Then he looked back down at the papers on his desk. Borman did not need to be told that he had been dismissed.

TWO

1962–64

THE PEOPLE AT NASA KNEW HOW TO PUT ON A GOOD SHOW, AND how to make that show irresistible to the press. The rockets were beautiful, the astronauts were dashing, and their families were charming. If you read all the stories in the newspapers and magazines, you couldn't help but believe that the space agency was showing and telling you everything you needed to know about how it went about its business. The space agency, however, was also good at keeping a very big secret: half the time their engineers were just making things up as they went along. No one with any sense would admit this part of the story to the public or the press, of course, much less to the U.S. Congress, which was providing the funding to keep the agency running. But from the beginning, a certain amount of on-the-fly improvisation had been inevitable, given that most of NASA's original recruits had come from Hampton, Virginia. And Hampton, as its residents knew well, was a town full of screwballs.

For generations, Hampton had been an ordinary enough place—a quiet community south of Newport News and north of Virginia Beach, with easy access to the waters of the Chesapeake Bay. All that changed early in the twentieth century, when the government started getting nervous about airplanes. After Orville and Wilbur Wright became the first people in the world to achieve powered flight, in 1903, the United States jumped out to an early lead in aviation and figured it was an edge the country could keep. But by 1915, America was getting clobbered by the Europeans. World War I had already been raging for a year, and if there was any good coming out of the slaughter, it was that the combatants were developing myriad ways to militarize aircraft for battlefield use. New technologies for fighting wars can easily be adapted to work in peace, and forward-looking U.S. officials realized that if they didn't act quickly, America might be left behind in the aviation business for good.

So Washington did what it so often does, which was to form an agency, the National Advisory Committee for Aeronautics. NACA's brief was to stay abreast of innovations in the field and make sure the United States regained its lead in aviation. At its inception, the agency was little more than a dozen men on a single committee representing the War Department, the Navy Department, the Weather Bureau, and others. It had only a single employee, an overall manager with an unremarkable background but an absolutely perfect name: John F. Victory.

NACA set up shop in Hampton, which was comfortably close to Washington and the War Department. The town also seemed like a good place for a research laboratory, which Congress was already planning to build. But if NACA started small, it grew quickly; by 1925 it had more than one hundred employees, with John F. Victory still overseeing most of them. The promised research facility, named the Langley Aeronautical Laboratory, got built, and over time NACA gobbled up other labs around the country as well. The agency played critical roles in developing American airpower for World War I, World War II, the Korean War, and in the push for supersonic flight that followed.

NACA's rapid growth meant that Hampton grew, too: by the 1950s, the airplane men who had settled there practically outnumbered the locals. And that's when things got odd. Practically every other person you saw on the street was one of the Brain Busters, as the locals had taken to calling them—a group of people who were uniformly male and uniformly young and uniformly walked around looking like their minds were on something else entirely.

The Brain Busters were the men who would go into a store to buy a washing machine or lawn mower and, before anyone could stop them, would lift the thing up or take off its back and begin poking around inside. They would ask the salesman how it was built and why this part had been used instead of that other part when any fool could see how much better the torque would be if only it had been designed right. The salesman—who had absolutely no idea how the machine was built— would agree, figuring that this response would close the sale or close the conversation, either of which was fine with him.

To the townspeople, the Brain Busters were often indistinguishable from one another, but among the airplane men themselves, one man stood out: a young engineer named Chris Kraft. Kraft had been born in 1924 in Phoebus, Virginia, which was less than two miles from Hampton, or nearly close enough to allow a young boy to smell the smartness on the air. Like a lot of the boys who lived in the area, Kraft decided he might like to give airplanes a try, and soon he was soaring through high school science and engineering. In 1940 he enrolled in Virginia Polytechnic Institute, and both classmates and professors there couldn't help but notice the compact, profane, and almost scarily smart young man, so tightly wound that he always seemed about to come unsprung.

When Kraft graduated from VPI, in December 1944, he wasn't particularly interested in working for NACA, though he applied for a job there because that's what VPI graduates did. What he really wanted to do was to work for Chance Vought, a private-sector aeronautics company in Bridgeport, Connecticut, that had money to spend and cachet to burn and was known for attracting the best and paying them accordingly. Kraft applied and was quickly hired, barely noticing in the excitement that NACA had also made him an offer.

He took a train from Virginia, through Manhattan and up to Bridgeport, and presented himself at the Chance Vought gate. Once he got there, however, the security people stopped him cold. Yes, they told him, he was on the list of new hires, but his paperwork was wrong or his birth certificate was missing or who knew what the problem was, but in any case, they wouldn't let him past the security checkpoint until it was corrected. He asked if he could speak to the man who had hired him; surely his new boss could straighten out the confusion. Rules were rules, they replied, and until the paperwork was in order, they would not talk to the personnel department. The best the company could offer was to put Kraft up in migrant-worker quarters outside the main gates until the next morning, when the problem would almost certainly be solved.

The next morning, Kraft again presented himself at the gate and again was told that there appeared to be some kind of paperwork problem. No, the guard insisted once more, he most certainly could not speak directly to the man who had hired him to resolve the confusion. After yet more delay, Kraft turned on his heel, strode to the nearest pay phone, and called NACA. He asked if the job was still open. It was.

That very day, he took a train back to Virginia, accepted the position, and wrote to Chance Vought, telling the company exactly what it could do with its private-sector job. When he filled out his new NACA employee forms, joining a group of engineers who could trace their organizational genes all the way back to a man known as John F. Victory, he allowed himself the flourish of signing his full and proper name: Christopher Columbus Kraft.

Kraft thrived at NACA. He earned $2,000 per year, which was more than sufficient, given the kind of work he was getting to do in exchange. But his newfound affluence and the joy he took in his job didn't make him any less ornery. Once he had put enough money aside, he and his wife, Betty Anne, bought a plot of land in Hampton with the intention of building a house. An engineer like Kraft was not about to trust the work of designing his home to some drop-in architect, however, so he did the job himself, taking special pride in the fireplace.

Having spent years figuring out how to make air move over the wings and body of an airplane—one that might be going faster than the speed of sound—he had little doubt that he could figure out the best way to make smoke go up a flue. His design for the brickwork was unusual—not at all like that of an ordinary fireplace—but he was certain that the smoke would flow up and out of the chimney without so much as a wisp ever leaking into his and Betty Anne's living room. Then he hired the bricklayers, showed them his design, and set them to work. A few days later, they had the fireplace built—and it looked nothing like what Kraft had ordered up.

"This isn't right," he told them.

"What's wrong with it?" the lead man asked.

"You didn't do it like I told you to do it."

"This is the way we build fireplaces all the time."

"Well, it's not the way you're going to build mine," Kraft snapped.

"It will work just fine," the man protested.

"Not fine enough," Kraft answered. "Tear the damn thing out."

The men did what they were told, and the house was left with a pile of bricks in the living room and a hole in the roof where the chimney should go. Finding another bricklayer willing to do the job took a while, but that was fine with Kraft. The wrong chimney would be far worse than no chimney at all.

☆ ☆ ☆

A dozen years after Kraft went to work for NACA, the agency suddenly became obsolete. On October 4, 1957, the Soviet Union successfully launched Sputnik, the first artificial satellite. Once again, a technological race had begun, and once again the United States was in danger of falling behind. Washington responded by creating a newer agency still—the National Aeronautics and Space Administration. And unlike NACA, this organization wouldn't begin its work from a standing start with a committee of twelve and one employee. Instead, NACA would be folded into NASA, and the new agency would be lavished with funding and told to go hiring.

One thing wouldn't change: the big thinkers at the old agency would become the big thinkers at the new one, and Kraft would be among them. But building machines that could carry men safely above the atmosphere was an order of magnitude harder than building machines that could merely carry men through it. Making the spacecraft themselves—the airtight pods that would be cockpit and home for the crew—was actually the easier part, if only because they would require little fuel to maneuver in space and were thus unlikely to blow up while they were still on Earth. The rockets themselves were another matter.

America's fleet little Vanguard booster, which was supposed to launch the country's first satellite into space just two months after the Russians launched Sputnik, instead exploded on the pad. Unlike the Sputnik launch, the Vanguard disaster was broadcast live, mortifying

NASA but galvanizing it, too. The agency's Atlas missiles, which most people considered the likeliest machines to put Americans in orbit, nonetheless inspired little faith, since they blew up about half the time they were launched. The smaller Redstone rocket, which would be used for America's first manned suborbital missions—little ballistic chip shots that sailed up into space and then fell straight back down to the ocean without ever entering orbit—were seen as scarcely better.

The Redstone was the direct genetic descendant of the V-2, the fat black-and-white missile the Germans had launched with such abandon against London during World War II. Both the V-2 and the Redstone were the brainchildren of NASA's chief designer, Wernher von Braun. He and much of his team of German engineers had been scooped up by the American Army after the war and offered new jobs building rockets for the United States—a very good deal, partly because the alternative was military arrest.

Kraft did not care a lick for the Redstone. Any rocket whose basic hardware had been designed to do its most important flying downward—to, say, Trafalgar Square—could not be fully trusted to fly upward into space. His skepticism only deepened when, early in his NASA tenure, he went down to Cape Canaveral to watch a test of a Redstone with a mock-up of the one-man Mercury capsule atop it. Since part of his job that day was to determine how the flight controllers and the launchpad crew worked together, he and most of the other people present were stuck in the windowless blockhouse that passed for Mission Control, rather than standing out in the open where they could actually see the bird fly.

Only two of the consoles in the launch room had television monitors showing the launchpad, and Kraft got one of them. When the countdown for the launch reached zero, the Redstone flared and roared just as it was supposed to, and the TV camera swept upward to follow the rocket's flight. Inside the blockhouse, the two monitors were filled with . . . nothing but empty sky. The camera panned back down, and there was the Redstone, having flown perhaps four inches before thinking better of it and settling back onto the pad, steaming and smoking.

The Mercury capsule, however, had not received the message that the Redstone had gone nowhere. Assuming the rocket was in flight, the capsule popped its little top and released the spray of confetti-like aluminum chaff that was supposed to help the radar technicians keep track of the Mercury. But now the chaff was fluttering about the Redstone like the remains of a giant, preposterous party favor.

And still, the Mercury wasn't finished. Sensing that the air pressure was now at sea level, the capsule concluded that it had completed its mission and was returning to Earth, so now it was time to deploy its parachute, which it did. The chute burst out in an orange-and-white billow, and as soon as it did, the Canaveral wind took hold of it, inflating it to its full width. With every gust, the parachute yanked hard against the rocket, which was still filled with tens of thousands of gallons of explosive liquid oxygen and ethyl alcohol.

"Is that chute going to pull the rocket over?" Kraft shouted into his headset to the engineers outside the blockhouse. The engineers answered—in German.

"Is-that-going-to-pull-the-rocket-over?" Kraft repeated, staccato this time.

More German came back.

"Would someone please talk to me?" he demanded. *"In English!"*

At last a few controllers Kraft could understand came on the line, but no one had any good ideas about how to proceed. Sending the rocket a command to open its valves and harmlessly vent its fuel might once have been possible, but it wasn't anymore, since the four-inch flight had unplugged the umbilical cord that had connected the rocket to the command center during its prelaunch phase.

"Can we plug it back in?" one controller in the blockhouse asked.

"How?" Kraft replied, already knowing the answer.

"Well . . . send someone out there to do it."

Kraft and the other men looked around the room at one another, then back at the two TV screens, where the fully armed Redstone bomb could be seen smoldering and swaying, the parachute still tugging at it with every gust of wind. Nobody volunteered for the job.

Rounding up a cherry picker, riding it out to the pad, and sending someone up with shears to cut the parachute's cords was an option, too. But the same lack of a volunteer with a death wish caused that vote to fail before it even reached the floor.

"We could shoot at it," came another suggestion from someone in the blockhouse.

"*Shoot at it?*" Kraft asked, his eyes wide.

"With a high-powered rifle. You know, to punch some holes and drain the fuel."

This idea was greeted with a thunderstruck silence, and the controller who'd offered it shrank back in his chair.

In the end, the men in Mission Control could do nothing but wait. The Redstone, they knew, ran on batteries, and batteries had only so much life. For hours they watched as the wind blew and the parachute snapped and the Redstone shuddered. Slowly the rocket lost more and more of its electrical life, until, finally, it went dead. When it did, all of its circuits returned to their safe position, its valves opened, and its fuel vented harmlessly away.

By the time the controllers had concluded that it was at last safe to go out to the pad and take down the dead rocket, Kraft—making his presence felt at NASA in much the same way he had at NACA—had established one of the most important of the space agency's growing list of flight rules: If you don't know what to do, don't do anything at all.

☆ ☆ ☆

Frank Borman and the rest of the second astronaut class were welcomed to NASA in 1962, and by that point the men who ran the agency had done a reasonably good job of setting their house right. Americans had flown in space four times now: Al Shepard's and Gus Grissom's suborbital missions, John Glenn's three-lap romp around the planet, and Scott Carpenter's duplicate three-orbit flight. All of these Mercury missions were triumphs—as long as you didn't look too closely.

Shepard's flight had indeed been flawless. Grissom's almost was, but then the spacecraft splashed down and the hatch blew too early and the astronaut almost drowned before being hauled out of the drink by a horse collar dangling from a helicopter. Glenn's flight, too, had been fine right until the end, when suddenly it looked like his heat shield—the thick plate of heat-dissipating material that would protect him from the 3,000-degree inferno of reentry—might be loose. This caused a lot of live-TV hand-wringing before the spacecraft at last hissed safely down into the ocean, with Glenn happy and unharmed. Carpenter's flight was the least successful of the four, though the mistakes were more or less of his own making. After faffing around too much with his orbital experiments, he hit his reentry rockets too late, causing him to overshoot his recovery area by 250 miles and requiring the Navy to go searching for him.

The press didn't make much of the problem, since a 250-mile manhunt in the open sea didn't seem very important when weighed against the great distances Carpenter had flown. And outwardly, NASA was perfectly happy to let reporters see things that way. Inwardly, however, several members of the NASA brass seethed, none more so than Kraft. By now he bore the title of chief of flight operations, and he would make it his business to be sure that those operations were conducted flawlessly.

"Never again," he said to the people closest to him in the flight office. "Scott Carpenter will never fly for me again."

Before telling other NASA administrators about his decision, however, Kraft dropped the hammer on Carpenter in person. That was only fair, and Kraft also gave the astronaut the dignity of not informing the press. But Carpenter was through, and he knew it better than anyone.

With the one-man Mercury spacecraft having achieved most of its objectives and only a couple of longer-duration missions left to fly, it was now time to roll out the bigger, more sophisticated two-man Gemini spacecraft, and to roll out, as well, the new astronauts who would fly them. But those men somehow seemed less exciting than the ships.

Americans had swooned when the Original Seven astronauts had been unveiled in 1959, and how could they not? The names alone—Gus and Deke and Gordo and Al, with their pleasing, off-the-tongue snap—had it all over the Soviets' Yuris and Ghermans and Pavels and Konstantins. Then there were the buzz cuts and the silver suits and the pretty wives, all of which seemed like the stuff of real explorers. So did the stories of the hard-drinking, skirt-chasing high jinks the first astronauts allowed themselves when they were away from those pretty wives, though the press did a good job of knowing and yet not knowing about what the men were up to. The public, meanwhile, did an excellent job of ignoring the rumors altogether.

A second crew of pilots could never be as celebrated as the first, and even a space-drunk nation knew it. Inside and outside NASA, the nine astronauts who followed the Original Seven were often known simply and glamorously as the Next Nine. Four of the new group—Frank Borman, Jim McDivitt, Tom Stafford, and Ed White—were Air Force, meaning that Borman's beloved service still hadn't achieved a majority. They were joined by four Navy men: John Young, Elliot See, Jim Lovell, and Pete Conrad. And finally there was an oddball, Neil Armstrong, a Navy man originally, and one who had flown seventy-eight combat missions over Korea. In 1960, however, he had retired and become a civilian test pilot. To some, it was more than a little awkward that NASA would recruit an astronaut who could dress only in civvies.

The Original Seven sniffed at their new colleagues, and the Next Nine acutely felt the disdain. The free Corvettes that auto dealers kept handing America's first astronauts would dry up fast if sixteen pilots were suddenly lining up for the keys. And the $500,000 deal the first astronaut class had struck with *Life* magazine for complete access to their training and their home lives wouldn't go nearly as far if the pie had to be cut into sixteen skinny slices.

But *Life* came through again, paying each of the new men $16,000 per year, an altogether princely sum for pilots who had spent a career earning military pay. And in its own way, NASA provided, too. When they first got to Houston in September 1962, the nine new astronauts and their young families were told to report to the Shamrock Hilton, a legendary place that advertised itself as "America's Magnificent Hotel." To all appearances, the Shamrock's nearly imperial lobby, massive ballrooms, and trapezoidal swimming pool with its three-tier diving board lived up to the hype.

Frank Borman was as dazzled by the place as the other astronauts were, but that didn't mean he was entirely comfortable with it. He and his family had lived modestly at Edwards, and they had driven from the air base to Houston in his 1960 Chevy. Just because his job had changed, he saw no reason to change his ways. Even a *Life* magazine contract didn't inspire him to begin throwing his money around.

"We can't afford this," Borman whispered to Susan when they walked into the hotel.

But as it turned out, they could. When Borman approached the desk and gave his name, the clerk glanced down at his guest registry and then looked back up, beaming at the man he had just learned was an astronaut. The clerk assured him that the Shamrock was at his disposal for as long as he and his family needed it. It was the least the hotel could do for an American hero.

No matter how glamorous the new lodgings were, Borman would accept them for only so long, especially since he had thus far done nothing to earn the adulation of the hotel's staff. As quickly as possible, he and Susan went house hunting. First they found a convenient rental, but before long they went looking for a piece of land on which they could build a home. They focused their search on the new communities popping up around town for NASA employees, all of which had names like Timber Cove and El Lago, though the bucolic-sounding silliness really had no place in hot and swampy Houston.

The Bormans settled on El Lago, picked a lot, and put down $26,500 for a construction contract. The house would be an extravagance, one that no pilot who served at Edwards could have imagined. But it was a comparatively modest investment, and after a dozen years of vagabond living and base housing, Borman had decided that Susan and their two boys deserved a real home.

☆ ☆ ☆

If any of the Next Nine were under the impression that an astronaut's job involved little more than learning to fly a spaceship, they were quickly disabused of that notion. The two-man Gemini hadn't even been fully built yet, the Titan booster that would carry it to orbit had not yet been man-rated, and a lot of work needed to be done before anyone would be taking any rides aboard either of them.

Yes, there would be plenty of basic training in centrifuges and simulators. The astronauts would spend endless hours in classrooms studying orbital mechanics, lifting bodies, and zero-g navigation, to say nothing of survival training on the ocean and in the deserts and anywhere else an errant landing like Carpenter's might deposit a pair of men. But that still left extra hours in the week, so all of the new recruits would be assigned a specialty, thus giving them a direct hand in developing the hardware, software, and flight procedures on which they would be banking their lives. The responsibility for matching up the man and the job went to Deke Slayton—and that, by any measure, was a very good thing.

Of all the Original Seven, Slayton had arguably been the astronaut the press had liked best. He didn't have John Glenn's charisma or Wally Schirra's twinkly wit, but he had craggy features, a no-nonsense air, and that wonderful blade of a single-syllable name. He was properly known as Donald Kent Slayton, but Donald Kent had been shortened to Donald K., which in turn had given way to D.K. and, ultimately, Deke.

The iron-man nickname, however, did not belong to someone with an iron man's heart. No sooner had Slayton been announced as one of the first astronauts than the doctors noticed a pattern they didn't like on his electrocardiogram tracings. There was an occasional wiggle, a sort of tremble—or a fibrillation, as they liked to call it.

More testing yielded the same result, so they grounded Slayton. Then, in a second indignity, they stripped him of his license to fly planes, lest he fibrillate in the cockpit and come crashing to Earth. Finally, having wrecked both his astronaut and his Air Force careers, NASA offered him a consolation job, one they decorously called coordinator of astronaut activities. The title sounded as if it belonged to little more than a camp counselor, and, in fact, the job had no clearly defined purpose. But it did mean that Slayton would get a paycheck and an office until he decided that he was ready to resign, quitting the space agency on more or less his own terms.

Slayton, however, had a better idea. Instead of fading quietly away, he decided to recast his job into something else: chief astronaut. Though he might not get to fly, he would make it his business to decide who did. By sheer force of personality, he pried away almost every bit of oversight the administrators in Washington, Houston, and Cape Canaveral had over the astronauts and gathered it to himself. Headquarters might have the final word in selecting the incoming astronauts, but Slayton would decide which of those men flew which missions and what work they would do in the meantime. And no mere administrator would get to any of the agency's astronauts without going through Deke Slayton first.

Slayton suspected that he would be good at the job, and he suspected right. If he felt bitter about his grounding, he never showed it. If he ever envied any of the men he was tapping to fly missions that he could have flown just as well, it was never evident. What *was* evident was that Slayton had a keen eye for talent and temperament: the head of the Astronaut Office had an uncanny ability to spot strengths in a man that the man himself did not know he had.

Slayton was impressed by Frank Borman the moment he met him, and what he liked most were the traits that were not unlike Slayton's own. He liked Borman's absence of flash, his ability to block out distractions, his sheer doggedness. He also liked the way Borman had fought back from his long-ago grounding; if Borman could overcome the ruling of the doctors, Slayton himself might, too.

Slayton gave the young astronaut a number of assignments, the most important of which was to investigate the Gemini's intended booster, the Titan, and make sure the rocket could fly without killing the men riding aboard it. Borman didn't know a thing about boosters—or at least he didn't know more than any new astronaut knew—but he was happy for the assignment, primarily because he had been suspicious of the Titan almost from the start.

The problem went to the very heart of the rocket's design. The simplest boosters had one main engine, but the Titan had two. Two engines meant two chances for something to go wrong, and that could be especially dangerous on the launchpad. If the rocket started to lift off and both engines quit, it wouldn't go anywhere. But the engines were mounted side by side, so if only one quit, the Titan would rise up and then go badly awry, flying first sideways and then down to an unsurvivable crash.

That scenario, Borman figured, was a risk that needed to be addressed. He dug into the Titan's technical manuals and manufacturing schematics, and although he saw some fail-safes that ought to prevent a one-engine ignition from happening, he was not impressed by what the design ought to do. All that mattered was what the engine *would* do, and something in his gut told him this rocket was trouble.

So, like Kraft before him, Borman went to see the rocket for himself, making it a point to be at Canaveral for the next Titan test launch. When he got there, he met the engineers who had designed and built the booster, and he raised his concern with them.

The lead engineer waved off Borman's worry. "It can't happen," he said.

"Anything can happen," Borman answered.

The engineer shook his head. "There could never be a failure mode in which just one barrel fires," he said flatly. "It's simply not the way the system works."

Borman let the matter go but stayed on the Cape for a few days and waited for the next Titan test. When the day for the new test arrived, he and the engineers crowded into the blockhouse. The rocket's systems were checked, the countdown clock marched toward zero, and the launch director called, "Ignition."

And on every screen—many more than just the two that had been used when Chris Kraft's sorry Redstone failed—what every man in the room saw was a single engine roaring with fire and a second engine remaining cold and dark and silent. The rocket strained to lift itself into the air.

"Shut down!" the launch director called, just before the crippled machine managed to leave the pad. That part, at least, worked as it was supposed to, and the Titan settled back to the ground.

Borman looked balefully at the nearest engineer, and the engineer—to his credit—met his eyes. The young astronaut had been right after all: *Anything can happen.*

☆ ☆ ☆

The first flight of the Gemini spacecraft, on April 8, 1964, was a lot less momentous than it might have been. Designed to be a shakedown cruise of the Titan booster and the Gemini itself, it would carry no astronauts into space. The ship would orbit the Earth three times, but it wouldn't even reenter the atmosphere under the parachute that would someday be needed to return a crew safely. Instead, it would simply burn up on the way down. And to ensure that no bits of debris would endanger anyone on the ground below or reveal too much about the spacecraft's design should parts of it rain down behind the Iron Curtain, NASA engineers drilled holes in the heat shield. The four-ton ship would be effectively vaporized.

The press showed little interest in a robot spacecraft that flew up and around, then came down and incinerated itself in the process. But when Gemini 1 accomplished those modest goals, NASA was elated. No American had been in space since Gordon Cooper's twenty-two-orbit, thirty-four-hour marathon almost a year earlier, and the United States was itching to be back in the game. After one more unmanned flight, Gemini 3 would at last take off and carry its history-making two-man crew into space. Beyond that mission, there would be up to ten more, with a Gemini due to be launched as frequently as once every seven or eight weeks. That meant a lot of seats for a lot of astronauts, and although one flight might seem to be as good as any other, the whispered word among all the pilots was simple: Whatever you do, don't get stuck with Gemini 7.

The problem with Gemini 7, as every astronaut knew, was that the mission apparently had but a single purpose: to torment its two-man crew for fourteen straight days and nights.

Ever since the Gemini had been first unveiled, there had been a lot of talk coming out of NASA about the grandeur of the new spacecraft. It would be so much bigger and so much more sophisticated than the Mercury ships. There was plenty of truth to that: the Mercury had been little more than a pod filled with instruments, with a single man stuffed inside. The pilot could fire his thrusters to waggle the ship this way and that. He could do a little sextant sighting and orient himself for reentry, though the computer could do that perfectly well if he allowed it to. He could also fire his retro-rockets to bring himself home, but the automatic system was more than capable of handling that job, too, for him. Even the Mercury's window was little more than a porthole over the astronaut's head, one he could see out of only if he craned his neck backward. For a pilot with any self-respect, the Mercury was as much carnival ride as spacecraft.

The Gemini, on the other hand, would be different. The astronauts flying it wouldn't be going to the moon, but it would be as close to a dress rehearsal for lunar flight as was possible.

The Gemini would have a proper cockpit, with two men sitting upright and side by side. Each would have a window directly in front of him that could be used to sight-fly the ship, the way a real pilot should.

The Gemini would be capable of rendezvousing and docking with other spacecraft, linking up in space the way an Apollo command module and a spidery lunar lander would have to do one day. The astronauts could fire their thrusters and raise their orbit to 800 or 900 or even 1,000 miles up, blowing past the 175-mile altitude record Wally Schirra had set on his six-orbit Mercury flight. And on at least a few of the missions, an astronaut would open his hatch and climb outside and walk—actually *walk*—in space, becoming, in effect, a human spacecraft hanging in the void outside a mechanical spacecraft.

The Gemini missions would do all that and more, but they would test another machine as well: the human machine. By NASA's calculations, the maximum length of a lunar voyage would be two weeks, far more time than any human being had ever dared spend in space. Before you could entrust your body to the mercies of the moon, you had to run the experiment much closer to home. If you were orbiting the Earth and your weightless blood began to pool in your brain or your heart forgot how to work after too much time in microgravity, you could fire your rockets and be back on the ground and in the arms of a medic within a couple of hours.

Somewhere in the middle of all those dazzling Gemini missions, then, there would have to be one long, gritty grind of a mission, one in which two astronauts would be sealed inside their ship, sent into space, and told that they would be allowed to return home only when fourteen days—or 224 orbits, or 336 hours—had been completed. No rendezvous or altitude records or spacewalking for these men. They wouldn't conduct experiments, or at least not many. Instead, they would *be* the experiment.

This was the mission every astronaut hoped to duck. For Borman, at least, the happy news that he would be spared the indignity of Gemini 7 came early.

Soon after word got out that the flight assignments were being made, Slayton called Borman into his office. "Frank," the chief astronaut said, "I've thought a lot about this, and I'm giving you Gemini 3. You're going to be first out of the gate."

Borman was elated. Gemini 3 would be a short flight—just three orbits, or a little over four and a half hours. If the mission took off early in the morning, the crew would be back on the deck of the recovery ship in time for a late lunch. Borman would be second-in-command of the two-man team, assigned to the subordinate's right-hand seat. The commander, on the left, would be Mercury veteran Gus Grissom. Yet Borman had no objection to playing a modest role on a modest flight. The historic significance of his selection was undeniable.

The names of the first seven astronauts were chiseled in the national consciousness like those of the founding fathers, and nothing could ever touch them. But now the next name on that list of explorers would apparently be Borman's. This was heady stuff, though what mattered more to Borman was that it was one more confirmation of the faith that Slayton, and, by extension, all of NASA, had in him.

Before the assignment could be considered official, however, Slayton told Borman that he needed to complete one more task. He had to go to Grissom's house and talk with the mission's commander face to face, pilot to pilot, so that Grissom could be sure he was comfortable with the man who had been tapped to fly with him. It was a condition Grissom himself had set, and Slayton agreed that it was a senior flier's prerogative.

Borman didn't know Grissom from sour apples, but he did know of his reputation. He was a first-rate pilot who had flown more than one hundred combat missions in Korea and had the tough-as-nails personality required for that kind of work. His friends and many of his colleagues liked his coarse and blunt ways. But to people who didn't know him well, he had much of the charm of a small stand of cactus.

Borman had dealt with far worse at West Point and in the Air Force, and he figured he could handle his conversation with Grissom well enough. He showed up at Grissom's house precisely on time, listened when he was supposed to listen, spoke when he was supposed to speak, even yes-sirred and no-sirred when those responses seemed called for. After the meeting was over, Borman felt reasonably confident that it had gone well.

The next day, Borman got another summons from Slayton.

"Frank," the chief astronaut said this time, "I'm taking you off 3. Gus wants John Young instead."

Borman said nothing; he didn't trust himself to speak. Slayton didn't explain why Grissom had requested the change, probably because

Grissom himself hadn't been required to explain. It was his ship, and he would pick his crew. Borman just nodded, thanked Slayton for his time, and left his office.

"Doesn't bother me," he would say in the days that followed to anyone who asked. "If he doesn't want me, I don't want him."

A few days later, Slayton called Borman into his office again, this time to inform him that he had drawn the short straw: he had been assigned to Gemini 7. The chief astronaut did offer one consolation, however: on this flight, he would be the commander. In the right-hand seat would be Navy man Jim Lovell, who, it seemed, had drawn a shorter straw still.

Borman and Lovell, as Slayton had ordered, promptly set about training for Gemini 7, the worst flight anyone could imagine—until it turned out to be something else entirely.

THREE

Summer through fall 1965

THE NEWLY NAMED CREW OF GEMINI 7 LEARNED A LOT OF THINGS while they were training for their first flight to space, but the most important of them was that no NASA doctor seemed to consider a week complete unless he got the chance to mess with an astronaut. To the medical experts, astronauts offered one of the greatest controlled experiments ever devised. They were a small group of hand-selected, exquisitely trained men, representatives of a species purpose-built to live in the one-g, radiation-shielded, temperature-controlled environment of Earth. Then those same men would be removed from that little incubator and hurled into a completely different environment for hours or days or even longer, while the entire scientific community studied how the men responded to such extreme otherworldliness.

Better still, the flying lab rats didn't even have to be coaxed to volunteer. Unlike graduate students lured with the promise of pocket money or academic credit, astronauts clamored for the job—competed fiercely for it, in fact. They would practically undertake it for free. It was all the doctors could do to contain themselves.

The first astronaut class had suffered seemingly endless testing, and it became an open secret that the medical examinations they had endured in order to be selected had been exhaustive, invasive, and flat-out humiliating. When the men were presented to the public at their first press conference in Washington, D.C., in 1959, they could have been forgiven if they had hobbled to the rostrum. At one point in the carefully stage-managed session, a reporter called out something to the NASA press spokesman who was fielding questions. The spokesman was clearly loath to repeat the query, but he could find no good way around it.

"The question is, would the, uh, gentlemen like to say, uh, which . . . which test they liked least?" the spokesman finally managed.

The reporters laughed, the NASA administrators squirmed, and the astronauts looked at one another and smiled. This was a question made for Wally Schirra.

Early on, Schirra had established himself as the bad boy and practical joker of the first astronaut class. It was Schirra who could never resist the opportunity—usually at a formal event with high-ranking Navy brass in attendance—to ask a fellow officer, in a voice too loud to miss, "Hey there. Are you a turtle?" Custom required that the man so asked would have two choices: he could either answer, "You bet your sweet ass I am" or say nothing at all, whereupon he would be required to buy a round of drinks for everyone in the room. Neither option was terribly appealing, but by the hard rules of the fliers' fraternity, choosing one was mandatory.

On the dais in 1959, Wally seemed prepared to answer the awkward question, but John Glenn stole the moment.

"That's a real tough one," he said. "It's rather difficult to pick one, because if you figure out how many openings there are on a human body and how far you can go in any one of them . . ."

Glenn trailed off and looked at the questioner with a nearly Wally-like twinkle. "Now you answer which one would be the toughest for *you*."

The room howled. Attractively and to good effect, Glenn blushed. Then he sat back in his chair.

Even Schirra was forced to concede the match. "I think he's answered for all of us," he said.

Now, in early 1965, with the flight of Gemini 7 in the planning stages, the doctors had the most promising lab specimens of all in Frank Borman and Jim Lovell, and they planned to make the most of them. Two weeks was an unheard-of amount of time in space, meaning the medics would need to establish a whole new baseline of data before they even got started. Borman and Lovell were made to repeat many of the awful exercises they had endured to get into the program in the first place—all the blood draws and dye injections, the electrified needles and centrifuge runs, the electroencephalograms and electrocardiograms and electromyograms.

Of particular and nearly obsessive interest to the doctors was the matter of calcium retention. Put a man in space for too long and his skeleton—which suddenly has little work to do, since it no longer needs to support his body against gravity—will stop wasting energy keeping itself strong. The calcium in the food he eats, which would ordinarily be used for maintaining and strengthening bones, will instead pass right through his body.

The experiment developed by the medical men was almost absurd in its thoroughness. The two astronauts would be required to save and bag every drop of urine and every gram of feces they produced in the nine days before the flight; do the same during the fourteen days aloft and for four additional days following; and turn it all over to the NASA doctors. Even that wouldn't provide enough data, however, so tears and sweat would also be collected before the mission. Further, Borman and Lovell would be required to stand in a wading pool in nothing but their skivvies while distilled water was poured over them, after which every drop would be collected and sampled for calcium. Before and after the flight, the medics would demand that the astronauts save and turn in their underwear—unwashed, please—at the end of every day so that it could be sampled for the sweat it had spent eighteen or so hours wicking up from the men's most personal places.

Blood pressure, balance, heart rate, respiration, and visual acuity would be tested as well, both before the flight and repeatedly while the astronauts were in orbit. But the vision test presented a problem. In a tiny two-man cockpit, it would be impossible to position an eye chart far enough away for it to be effective. Even if it were possible, the doctors wondered if the men would run the test honestly and tell the truth if their vision was going soft, or if they would lie to avoid the risk of being told to come home early.

So no eye chart would be carried aboard. Instead, on a vacant plot of land forty miles north of Laredo, Texas, NASA groundsmen would flatten and rake eight squares of terrain—two thousand feet long to a side each—and cover portions of the squares with either white Styrofoam or dark turf. The astronauts would have to describe the pattern of the alternating dark-light squares as they flew overhead, a pattern that could be switched up every time the spacecraft passed over Laredo. Let the flyboys try to cheat on *that* one.

And then finally the doctors went too far. "If you're going to try staying in space for two weeks," one of the doctors said, "it's probably smart to simulate it on Earth first."

"We're spending half our time in simulators as it is," Borman answered.

"But only for a few hours at a time," the doctor said.

"We could always book more," Lovell said, as Borman nodded in agreement.

"Right," the doctor responded. "But what we're talking about is simulating the whole thing, the whole two weeks—beginning to end. Just to play it safe."

Borman was incredulous. "You want us to spend fourteen days in straight-up seats? In a one-g environment—with no bathroom breaks?"

"Well, yes," the doctor said.

"Are you out of your *minds*?" Borman barked.

The doctors had no good answer for that, though they surely would have offered one if they did. At the prerogative of the mission commander, that experiment was duly scrapped.

If anything made the notion of spending two weeks in the corrugated can that was Gemini 7 more tolerable to Borman, it was the prospect of going aloft with his fellow astronaut Jim Lovell.

Borman had met Lovell during the astronaut selection process, when they and a group of other applicants were undergoing physicals at Brooks Air Force Base, in San Antonio, and he'd liked him from the start. For one thing, there was Lovell's temperament. The military was full of growling types like Grissom and joker types like Schirra and, Borman had to admit, hard-driven, hard-charging grinds like himself. What it was short on were the easygoing, uncomplicated types, the men who never got rattled. They stayed cool not because they didn't take things seriously or failed to understand the odds but because it was simply not in their nature to run hot.

"If you can't get along with Lovell," went the popular wisdom in the astronaut corps, "you can't get along with anyone."

Just as appealing to Borman was Lovell's background, which was so much like his own. He could forgive Lovell's choice of the U.S. Naval Academy over West Point because, as any West Pointer would tell you, every man is entitled to one grievous act of folly in his life. But Lovell, just like Borman, had had to scrap his way into his school of choice. He, too, had been given a third-alternate slot at first, but in his case the long shot hadn't come in and he'd spent two years at the University of Wisconsin before reapplying to Annapolis. This time he'd been accepted, though the naval academy required that he start over and begin as a plebe, which, he'd decided, was worth it if an Annapolis degree and a naval commission lay at the end of the line.

Lovell's path to NASA had been even more serpentine than the one he had followed to Annapolis and, again like Borman, he, too, had a tale about Gus Grissom to tell. Lovell applied to NASA in the first round of astronaut selection in 1958, and he survived deep into the vetting process, making it as far as the final group of thirty-two candidates before being bumped on a medical technicality—a high level of bilirubin in his blood. Even the doctor who gave him the bad news admitted that the anomaly was indicative of nothing terribly meaningful, but there were thirty-one other men still standing who didn't have that abnormal reading and only seven slots to fill. The field had to be winnowed somehow.

The medical exams had taken place at the Lovelace Clinic in Santa Fe, New Mexico, and Lovell returned to his quarters to pack his bags before flying back to Naval Air Station in Patuxent River, Maryland, where he was stationed. The astronaut candidates who survived this round would be going on to Wright-Patterson Air Force Base, in Ohio, for more tests. Before Lovell could leave Lovelace, however, he got word that he had a long-distance call from his wife, Marilyn, who was on the phone in the base office. Reluctant to give her the bad news, he walked to the office slowly. But before he could even get out what he had to say, she cut him off.

"Don't come home," she said.

"What?"

"Don't come home. NASA's been calling here all day; you're supposed to go straight to Wright-Pat for more tests."

"That's not what the doctors here said."

"It's what the man on the phone said. He wants you to leave right now and check into the officers' quarters there by tonight."

Lovell, faced with two sets of contradictory orders, did what any sensible officer would do in that situation, which was to choose the one that suited him best. He caught a flight to Wright-Pat, checked in as ordered, and immediately began scouting around. Knowing that NASA was sending seven men at a time for the tests, he immediately began looking for familiar faces and counting heads. He recognized one and then another and another of the men from Lovelace, but the count stopped at six— meaning he must be the seventh. Clearly, this was not a mix-up. Whatever the Lovelace doctors' worries about his bilirubin, they had obviously been overruled by someone higher up the military chain.

Lovell ate a happy breakfast the next morning, speculating with the other six men about what the Wright-Pat doctors could possibly do to them that the Lovelace doctors hadn't already. And then, uninvited— and, to Lovell's mind, deeply unwanted—a vaguely familiar Air Force man appeared at the mess hall door and made straight for their table.

"Grissom," the man growled, introducing himself to the group and shaking the nearest hands. "Sorry I'm late. Transport problems."

Eight men now sat at the breakfast table, and all eight recognized the awkwardness. Lovell forced the rest of his breakfast down and then went to see the NASA team at the base hospital. There he was told that yes, his orders to come here were in error; no, he had not made the cut after all—his bilirubin had indeed disqualified him. Grissom or one of the others had been the unknowing beneficiary of Lovell's misfortune.

Lovell went home as ordered, but as he had done with his initial rejection from the naval academy, he took the no as a temporary answer, one that would be set right in the future. In 1962, when NASA was screening its second class of astronaut candidates, that no turned into a yes.

Just as appealing was the story behind Lovell's marriage to Marilyn. Like Borman, Lovell had met the girl he would marry when they were both in high school. Also like Borman, Lovell had departed for his military education after having made the promise of marriage upon graduation. But unlike Borman, Lovell hadn't almost screwed the whole thing up with an impulsive decision to break off the romance during his first year away.

While Lovell attended the naval academy, Marilyn Gerlach, as she was then known, was studying liberal arts at George Washington University. She came to Annapolis nearly every weekend, as did a lot of the other midshipmen's girlfriends—or "drags," as they were called in academy argot, a nod to their roles as an upperclassman's attractive appurtenance. When visiting Annapolis, Marilyn stayed in one of the local rooming houses, a tidy multistory home run by a woman she and the other girls addressed only as Ma Chestnut. If the landlady had a different first name, nobody knew it.

One afternoon, in Lovell's last year at the academy and his and Marilyn's seventh year as a couple, they went for a walk in downtown Annapolis and happened to come upon a jewelry store. As they both knew, midshipmen visited this shop to buy their drags the miniature ladies' version of the naval academy ring, the one with a diamond in the middle, that would also serve as an engagement ring for those soon-to-be officers who decided it was time to wed.

Standing at the jewelry store's window, Jim looked at Marilyn and said tentatively, "So . . . choose whichever one you like."

"Whichever one I like?" Marilyn asked, either not understanding what was being offered or affecting not to understand, but either way flummoxing Lovell nicely.

"You know," he said, "as your engagement ring."

"Engagement ring? Just like that?"

"Well, I just kind of assumed, after seven years . . ."

"You just kind of *assumed*?" Marilyn answered. "James Lovell, do not assume anything at all unless you ask me properly."

And so, right there on that sidewalk, the young midshipman dropped to one knee and did ask properly, and Marilyn said yes. As the wisdom went, if you couldn't get along with Jim Lovell . . .

☆ ☆ ☆

Before Gemini 7 could get its chance to go to space, Gemini 6 would have to fly. Geminis 3, 4, and 5 had been almost complete successes, with the high point being Ed White's Gemini 4 space walk, the first ever for an American astronaut and a powerful statement about the rising prestige of the Next Nine class. The flight of Gemini 6 would begin with a launch less than two months before Borman and Lovell's scheduled liftoff in December 1965. Both men planned to be on hand to watch the Gemini 6 crew—Wally Schirra and rookie Tom Stafford—take off.

October 25 broke as a perfectly clear morning in coastal Florida. The day was unusually hot for so late in the year, but that didn't stop the tens of thousands of campers and spectators from lining the beaches surrounding the launch site, and it didn't stop the big three television networks and all the major newspapers from converging, too.

Swarms of people and press had attended the Mercury launches, and the same was true for the first three Geminis. But Gemini 6 drew easily one of the biggest crowds yet, mostly because there would be two launches that morning, barely ninety minutes apart, instead of the usual one. For the gawkers, that meant twice the number of countdowns and twice the level of thrills. For the morbid—and the media—it meant twice the chance that something would go disastrously wrong.

Borman and Lovell, watching from the VIP stands, weren't thinking about what might go wrong. They knew the risks in a personal and professional way. They understood that one or both of the rockets might come to ruin, or they might not. That was simply the binary way the game was played. Today, the two astronauts were pleased to take a little time away from their sixteen-hour days of training and watch the show like anyone else.

Off to their left was launchpad 14, where an Atlas rocket bearing an Agena spacecraft was standing fully fueled, venting liquid oxygen vapor, preparing to blast off to a neat, circular orbit 213 miles up. No one was aboard that rocket, and therefore fewer eyes and fewer TV cameras were turned its way, even though it would be the first missile to be sent flying today—in just half an hour, in fact.

More important was the machine on pad 19, the pad rated for manned launches, which was a good six thousand feet away from pad 14 and farther still from the VIP stands. On that pad stood a bigger and far more important booster, this one a Titan with a Gemini spacecraft perched on top and Schirra and Stafford tucked inside. The second launch would be the day's true headline event.

If everything went exactly as NASA had scripted it, the Atlas would light at 10:00 a.m. Less than a dozen minutes later, the missile would place the Agena—a twenty-six-foot-long robot vehicle with a rocket engine at one end and a docking port at the other—into its orbit, and it would take just ninety minutes for the spacecraft to complete its first circuit of the Earth. Shortly after 11:30, when the Agena passed back over the Cape, Schirra and Stafford would follow it into space. The two astronauts would spend a day chasing the Agena down and then park the nose of the Gemini into the collar of the Agena, locking them together like a single vehicle.

NASA's moon landing plans called for exactly this kind of maneuver: someday, a lunar excursion module, or LEM—which the engineers liked to pronounce as a crisp, single-syllable "lem," rather than as initials— would lift off from the surface of the moon with two men aboard. It would dock with a command module, with one man aboard, that was orbiting overhead. Once all three men were safely inside the command module, they would jettison the LEM and all three would come home. The United States had never managed such a rendezvous and docking feat even in Earth orbit, and the Soviet Union hadn't, either. This would be a good chance to practice the maneuver, check a box in the pre–moon mission to-do list, and tweak the Russians as well. The entire Gemini 6 mission would take just two days, but the space race would be transformed after it was over.

So far, things looked promising. "There hasn't been the slightest delay on the countdown of the Atlas-Agena rocket," CBS launchpad correspondent Charles von Fremd reported to Walter Cronkite, who was anchoring his coverage of the mission a safe three miles from the launch site, where all of the networks had set up their remote studios. "As of now, the countdown is going clickety-clack." Reassuring any viewers who still harbored worries about the flightworthiness of the Atlas—an interconti-nental ballistic missile with a habit of blowing up on liftoff—von Fremd pointed out that the rocket "is batting twenty-eight for thirty in its most recent launches."

With just over two minutes to go, the TV cameras switched away from the correspondents to a view of the Atlas-Agena on the pad. A digi-tal clock superimposed on the screen counted down the seconds; pre-cisely at zero, the engine of the unmanned rocket lit.

"Liftoff right on the button, right on the hour," Cronkite exclaimed. "There goes that fiery boom of the Atlas."

Cronkite, who had been broadcasting liftoffs since the very begin-ning, had a sweet spot for the temperamental missile that had gotten American astronauts off the mat and into orbit after the Soviets had taken the lead in the space race. The anchorman's enthusiasm for the Atlas always showed when one of them flew.

But just under five minutes into the flight, at the moment the Agena was supposed to separate from the booster, the tracking station on the Canary Islands reported that it had lost the signal from the craft, with data screens going blank both there and in Mission Control. What then took the place of the healthy vital signs of a spacecraft speeding to orbit was the worst thing possible: ground-based radar stations were no lon-ger detecting a single, upward-moving reflection but five scattered reflec-tions spreading across the sky in what could only suggest an explosion.

"That's not necessarily fatal," said Paul Haney, NASA's mission broadcaster, whose voice intercut with Cronkite's.

But of course it was fatal. The tens of thousands of people who had camped on the beaches to watch the two liftoffs might not have under-stood that, but the people in the VIP stands—at least the ones who worked for NASA—did. Schirra and Stafford, strapped into their rocket ship to nowhere, knew it, too. And so did every man at every console in the launch control center.

It would be a full fifty minutes before the Canary Islands tracking station confirmed the death of the Agena and, by extension, the scrub-bing of the Gemini 6 launch with the simple words "No joy, no joy." By then Borman and Lovell had already quit the viewing stand for the launch control center and found the predicted gloom. They also, however, found the beginnings of a mad bit of improvisational genius.

Chris Kraft was huddled with Walter Burke—a chief engineer with McDonnell Aircraft, the builder of the Titan booster—and Burke's assistant, John Yardley. The three men were surrounded by a gaggle of other NASA and McDonnell officials. All of the people there knew the engineering problems and the calendar math they were up against. It could take as much as six months to configure a new Agena and mate it to an Atlas, which would mean allowing Gemini 6 to slip while other missions moved up to take its place. Or NASA could put the brakes on the entire manned program until Gemini 6 was ready to fly, but with Gemini 7's December launch just forty days away and Gemini 8's launch only fifteen weeks later, the missions would begin piling up fast. And there was no telling—as there was never any telling—just what the Soviets might have cooking during that slowdown.

Standing in the middle of the small mob that now included Borman and Lovell, Yardley gave voice to what a few others were already thinking.

"Suppose we use a Gemini as a target instead of an Agena?" he asked. For those who didn't quite understand his point, he added, "Launch Gemini 7 first and make it the target for 6."

"You're out of your mind," Kraft said immediately. "That can't be done." But even as he spoke the words, it was becoming clear to him and everyone else in the room that it could indeed be done. In truth, it was practically begging to be done.

A joint mission would mean four men in orbit at once, placed there by a space program that only four years ago had struggled to launch even a single man on a suborbital high jump for a five-minute glimpse of space. Four men with four families on Earth had it all over the original plans for today—two men flying tandem with a faceless machine like the Agena.

Once the two Geminis were in orbit, there could be no docking between them; the ships weren't built that way. But it was the rendezvous part of the mission that was the hardest and most important: two pickup truck–sized vehicles that were moving at 17,500 miles per hour would have to find one another in millions of cubic miles of near-Earth space and then approach to within just inches of each other. The astronauts would get so close they'd be able to recognize one another's faces in the windows; achieving that kind of precision would be a big step forward for the space program.

Launching a dual mission would not mean moving up the date of the Gemini 7 launch, not least because Borman and Lovell still had a month of training to go. But it would mean sending the two spacecraft up in reverse order. Once Gemini 7 launched and was in orbit, the ground team would have eight to ten days to roll out Gemini 6 and get it into space while a few days remained in the Gemini 7's two-week mission. That scenario was possible, but it would call for some corner cutting, especially with only one man-rated launchpad.

It usually took at least a month to perform the repairs and maintenance that any pad needed after it went through the controlled firestorm that was a liftoff. That maintenance, however, also included cleaning and painting—which did nothing to improve the pad's performance but was important to reassure the frequent congressional visitors that Washington's money was being well spent down in Florida. This time, NASA would dispense with the cosmetic work.

Time could also be saved if the two spacecraft switched boosters. Gemini 6's Titan was already on the pad and had already cleared its preflight fitness tests. It could simply be left where it was, and the Gemini 7 spacecraft—which was equipped with more robust batteries and life-support systems for its longer stay in space—could be trucked out to the rocket and dropped on top of it. Gemini 6 would go back into the hangar and be positioned atop 7's Titan.

While Kraft and the others continued discussing the feasibility of this improvised plan, two of the four astronauts who would be sent on the new mission were still out on pad 19. By now they were probably riding down the gantry elevator, still wearing the space suits that had been intended for orbital space but were all wrong for the Florida sun. They would have to join the discussion later.

The two astronauts in the launch center, however, didn't hesitate to express their feelings about the plan.

"I like it," Borman said emphatically.

"It's smart," Lovell added.

Despite himself, Kraft had to admit that he was warming to the makeshift madness of the thing. NASA would have to resolve a number of problems first—how to modify a global tracking web built to follow one spacecraft so that it could keep track of two; whether the navigational algorithms could be written in time for the computers that would help guide the two spacecraft. But those were the kinds of problems Brain Busters had always been paid to solve, and these were the kinds of made-up, on-the-fly missions astronauts always hoped to participate in. Kraft had little doubt that when Schirra and Stafford heard the plan they would be just as enthusiastic as Borman and Lovell.

☆ ☆ ☆

The Agena spacecraft that was supposed to go into orbit on October 25 had been dead for just seventy-two hours when Bill Moyers took the rostrum at a press conference at President Lyndon Johnson's ranch in Austin, Texas. Only thirty-one years old, Moyers had not been Johnson's first official spokesman or even his second. Johnson went through press secretaries fast, but he liked Moyers a lot. He was young, frank, and a Texan, like Johnson himself. Moyers was also one of only fifteen people who were visible in the birth-trauma photo of the Johnson administration—the searing picture of the new president taking the oath of office aboard Air Force One just under two years ago, standing next to the shattered former first lady barely two hours after her husband had been murdered. Moyers was an indelible presence that day, and in one way or another, he had been at Johnson's side ever since.

There had been a lot of fretting within the space community when the Kennedy administration had suddenly become the Johnson administration. Kennedy had made the bold promise to get to the moon before 1970, and many at NASA doubted that the sleepy-looking new president could ever match the visionary energy of the previous one. But appearances did not tell the whole story.

It was actually Johnson who, as majority leader of the Senate, had bulled through the National Aeronautics and Space Act of 1958, creating NASA in the first place. And it was Johnson who, in his last weeks as a senator following his and Kennedy's election, had quietly engineered amendments to the law, so that most of the passages that vested this or that bit of executive authority in the president of the United States were changed to add the words "or the Vice President of the United States." Just six days after Kennedy's death, Johnson announced in a televised speech that the name of NASA's Cape Canaveral would be changed to Cape Kennedy. The name change offered powerful reassurance to voters and taxpayers that the new president would personally make certain that Kennedy's lunar dare would be honored.

Today Johnson's press secretary would surprise the assembled journalists: he had some real news to share, and it would be news about space, which was the reporters' favorite kind. Moyers began the morning's gathering the way he began many of them when the White House set up temporary operations at the ranch—with a description of the president's comparatively quiet day so far.

"The president was up at six o'clock this morning," he reported. "He spent some time clearing his desk of paperwork. He dictated some memoranda, some letters, and he and Mrs. Johnson walked down the road— between three and four miles, round-trip."

"What road are you talking about, Bill?" a reporter asked.

"The dirt road in front of his ranch," Moyers answered.

The press took down that little detail, which was the only scrap of color so far.

"While at his desk," Moyers went on, "the president talked to Secretary McNamara on numerous defense matters, particularly as they relate to South Vietnam."

Not much news here, either. The president seemed always to be talking to McNamara these days, and always about Vietnam. The reporters looked to Moyers for more, and the press secretary took mercy on them.

"The president went over some material on the space situation, and I am giving you a memorandum to the president from James Webb, administrator of the National Aeronautics and Space Administration, to the effect that NASA is going to try, while Gemini 7 is in flight, to launch Gemini 6 and have the two manned rockets fly in formation."

That was news, big news. Never mind that Moyers had called the spacecraft "rockets," even though by the time the two Geminis got into orbit, the Titan rockets that had carried them there would have already fallen away into the ocean. The reporters who knew something about space would use the proper language in their stories; the ones who didn't would be corrected by the editors who cleaned up their copy.

"Is this new?" a reporter called out.

"This is brand-new."

"It has not been announced before?"

"No."

"Will 6 be manned?"

"Six will be manned with two astronauts—the same who were just scheduled to go up."

"And 7?" someone pressed, either not having listened carefully or not believing what he'd heard.

"Right," Moyers said patiently. Then, spelling it out, he said, "There will be four men in space at the same time."

He then gave the press the four astronauts' names, ages, ranks, hometowns, and flight status—all but Schirra would be first-timers in space. In three different ways, Moyers confirmed that yes, the two spacecraft would be flying in formation; no, it was not possible to say exactly how close they would get ("a matter of feet"); and no, there would most assuredly be no docking attempted ("They will not touch").

"Bill, can we ask for a break of about five minutes to file?" a reporter requested.

The young press secretary nodded; they could have as much time as they wanted.

"This is all that I have," Moyers said. Then he stepped away from the rostrum and let the reporters do their work.

FOUR

December 4, 1965

THE PRECISE MOMENT WHEN A TITAN ROCKET IGNITED UNDERNEATH an astronaut for the first time always came as a surprise. The pilots all thought they were prepared for it, because the people who trained them had sworn they would be. But when the moment finally happened, the astronauts realized that the trainers had no idea what they were talking about.

The first thing no one mentioned was the glug-glug-glugging sound that began thirty seconds before the engines even lit. More than 120,000 gallons of two different kinds of hot-tempered fuel were sloshing around in the 103-foot booster, and if the rocket was going to go anywhere at all, those volatile chemicals would have to flow and mix. And since the fuel pumps were high up in the booster, right beneath the Gemini spacecraft perched on top, the sound—like that of a giant bathtub draining—was impossible to miss and impossible to feel terribly good about.

The astronauts also weren't prepared for the way the booster would sway in the wind while it was waiting to take off. The higher up the stack, the worse the sway, which meant it was worst of all for the two men tucked inside the Gemini at the very top. And though the astronauts had spent hours in centrifuges getting used to the seven g's they would be pulling after they at last took off, they didn't fully appreciate how quickly they would feel the grip of that force, almost the moment the rocket jumped off the pad. That was yet another feature of the Titan's ballistic missile pedigree: a weapon that's taking to the sky to defend the homeland from attack can hardly afford to dawdle while getting into the air.

Most of all, there was the noise of ignition. During the endless simulations, the trainers had never bothered trying to re-create it, because even the best sim could not reproduce the cannon roar of a Titan's engines. You had to be inside the cannon itself, which is exactly where an astronaut would be on launch day and at no other time. Without the help of the radio, it would be hopeless to try to talk to the man in the next seat; the closed helmet muffled the astronauts' voices, and the din of the rocket drowned everything else out. Even with the helmet microphone positioned just an inch from their mouths, the astronauts still had to shout to be heard.

At 2:30 p.m. on December 4, 1965—precisely the date and time the NASA flight controllers had aimed for, even with all the improvisational planning they'd done in the past six weeks—Frank Borman and Jim Lovell experienced the unimaginably ferocious blast of a Titan rocket for the first time.

"We're on our way, Frank!" Lovell yelled to Borman.

Borman, rattling about in the left-hand seat, said something or other in response, but it was hopelessly lost in the noise, even to his own ears. He tried again, this time settling for a quick, affirmative "Right!"

"Communications are a little noisy from the spacecraft," said NASA commentator Paul Haney, understating the reality by a considerable factor for the tens of millions of people watching on television.

The days leading up to the launch had been predictably busy, not only for Borman and Lovell but also for their understudies, Ed White and a rookie from the new third astronaut class, Mike Collins. White and Collins were the backup crew, and both men had to be ready to fly on very short notice.

Borman liked Collins, who would replace Lovell in the right-hand seat if circumstances made that necessary. He found Collins exceedingly smart and engaging, with a dry but lacerating wit and a near-lyrical gift for language. It was also hard not to notice Collins's perceptive take on the people around him. Much later, Borman learned that Collins had sized him up as a driver, a capable and aggressive man with the habit of carrying himself like a politician who had an election approaching. In this case, as in most cases, Collins was spot-on.

But Borman found it especially fitting that his own understudy was Ed White. Borman had been a military man and an astronaut long enough to know how to establish a camaraderie with the men around him. But camaraderie and abiding friendship were different things, and though he shared the first with a great many people, he shared the second most powerfully with White. The Bormans and the Whites lived across the street from each other, and the families got along unusually well. Susan and Pat were closer than nearly any other pair of women in the astronaut wives sorority, and the Borman boys—Fred and Ed—got on well with the White children, Edwin and Bonnie Lynn, with admirably little teasing directed at the one girl from the three boys. The four friends and their children often spent weekends together, and the two astronauts would take the opportunity to fish and talk space, or fish and talk about anything other than space.

On Gemini 7's launch day, Borman, Lovell, and their two backups awoke at 7:00 a.m. Both crews prepared meticulously for liftoff; before Borman and Lovell even entered the capsule, White and Collins spent an hour there, checking out the systems and making sure the craft was ready to fly. Just before noon, the two second-stringers climbed out and the two primary astronauts were loaded into the spacecraft and strapped into their seats. Then the twin hatches—one over each astronaut's head—were closed.

The hours before ignition moved exceedingly slowly. Borman was glad that he and Lovell had eaten a big breakfast that morning; the meal was the last good one they would have for a while. Plenty of fine-sounding food was stowed aboard the spacecraft for the fourteen days they would spend aloft—chicken and vegetables, shrimp cocktail, beef and gravy, butterscotch pudding, packaged fruits—but all of it had been shrink-wrapped or freeze-dried to within an inch of its life, and all of it would have to be eaten with a spoon from a plastic pouch that had to be opened with a pair of surgical scissors. There would also be fruitcake—lots and lots of fruitcake—packaged like unholy sausage links in a long necklace stowed behind Lovell's seat. It was energy-dense and high in calories, though nobody pretended it would taste any better in space than it did on Earth.

But now, as their rocket climbed and the g-forces built, food was not remotely on the astronauts' minds. At two minutes and thirty-nine seconds into the launch, exactly on schedule, the first stage shut down and fell away, flinging Borman and Lovell forward against their seat restraints. A moment later, the second stage lit, slamming them back in their seats. The g-forces climbed on schedule to four and then five, six, and seven, meaning that within seconds the two men, who weighed the astronaut average of about 155 pounds on Earth, clocked in at nearly 1,100 pounds on the balance scale of gravitational physics.

And then, five minutes and forty seconds after they left the coast of Florida—less time than it would have taken to walk from the launchpad to the Cape Kennedy commissary—their engine cut off, their spacecraft entered orbit, and the astronauts suddenly weighed nothing at all. Around them, bits of dust and the odd screw or bolt, the sorts of things inevitably left behind in even the most painstakingly prepared spacecraft, floated lazily in the air. Like all first-timers, Borman and Lovell poked at the drifting flotsam, grinned at it, and then grinned at each other.

"Gemini 7, you are go!" called up the capsule communicator—or capcom—in Houston.

"Roger," Borman answered, "thank you."

"That's the best sim we've had so far," the capcom joked.

Borman and Lovell, who now formally and forever had joined the small fraternity of men who had flown in space, simply grinned again.

☆ ☆ ☆

It didn't take long before Gemini 7 became precisely the 336-hour grind it had been advertised to be. For a fortnight, a pair of full-grown men would be locked in an enclosure with no more habitable volume than the front seat of a Volkswagen Beetle and even less room overhead than the Beetle offers. The two astronauts quickly learned that they could extend and stretch their legs, but only if they bent their upper body. Or they could stretch their upper body, but only if they bent their legs.

Doing both at once was impossible. There was also a lot of instrument noise, the whirring and ticking and whooshing of the thrusters and the ventilators that were the very heartbeat of the ship. Borman and Lovell might have found the noise reassuring, except that it never, ever stopped.

Even sleeping wasn't easy, with the capcom's chatter added to that constant background sound. That back-and-forth with the ground didn't stop either, because NASA rules required that at least one astronaut remain awake at all times, and polite whispering was nothing like silence when your fellow pilot sat just inches away from you.

The menu, meanwhile, lived down to its advance billing. Foods that had to be rehydrated with a water gun never achieved the right texture: dry, powdery bites alternated with sticky, watery ones, and no single mouthful struck the proper balance. The fruits packed with sugary syrup were a little better, as was the punch. The fruitcake was . . . fruitcake.

Borman and Lovell kept as busy as they could with medical experiments and navigational sightings. They worked with the Navy to track Polaris missiles launched from deep-ocean submarines, a job that was both challenging and distracting, just as the NASA psychologists had hoped. Unavoidably, however, there was a lot of idle time. Both men had brought along a book: for Borman, it was Mark Twain's *Roughing It*; for Lovell, *Drums Along the Mohawk*, the 1936 Walter Edmonds best seller. The choices pleased NASA; they were exactly the wholesome fare the agency wanted the world to see its astronauts reading. But in reality, neither book got more than a glimpse, though now and then Lovell did make a few notes in a journal he had packed.

The space suits themselves became impossible. For a fourteen-day flight, the usual fighter pilot pressure suits and the hard-shell helmets simply would not do, so NASA had ordered up lighter, softer suits with cloth helmets that zipped open and closed and could be folded back like a parka hood. The astronauts and most other people at NASA promptly dubbed them get-me-down suits, since they were not usable for space walks or much of anything beyond allowing the pilots to survive a sudden depressurization on takeoff or reentry. They were also clingy and almost unbearably hot. Peeling them off would leave the crew in their far more comfortable long johns, but NASA didn't like that idea, since get-me-down suits can't actually get you down if you're not wearing them.

So Borman sweltered in his suit, turning his air circulation knob to its coolest setting, which helped only a little, while Lovell, over the course of the first few days, slipped slowly out of his. First he eased the suit down around his shoulders; surely NASA wouldn't object to that. Then he lowered it to his waist, and finally he freed all but his lower legs. Both men reported their discomfort to the ground, and over the next six days, the question of suits or skivvies rose up the chain of command. The capcom passed it on to the flight director, who passed it on to Chris Kraft, who passed it on to the deputy administrator in Washington, who consulted with the lead flight surgeon, who reported that yes, the biomedical readings he was getting from the ship showed that Lovell— who by now was fooling no one with his secretly vanishing space suit— had healthier blood pressure and pulse readings than did poor suited Borman. The word thus came back down the chain that it was the opinion of the NASA brass that the advantages of flying without suits outweighed the advantages of wearing them, and the men of Gemini 7 were officially cleared to fly in their underwear.

As the first day unfolded into the first week and the unshowered, unshaven astronauts grew grittier and riper, there was the ongoing problem of how a man preserves his last scrap of privacy—to say nothing of dignity—in a spacecraft with no proper toilet facilities. Urinating in space was not a problem, and the doctors, it turned out, had decided that they would need only occasional samples. This meant the men could usually relieve themselves into a tube and then vent it through a small port on the exterior of the spacecraft, where it would instantly burst into a shower of glittery crystals, a phenomenon Schirra had dubbed the constellation Urion.

But urinating wasn't the only problem, and the remaining one required plastic bags and disposable wipes and a lot of maneuvering if an astronaut was to take care of the matter properly. The other fellow, meanwhile, either pretended not to notice or acknowledged that he very much did notice and asked what in the world the man next to him had eaten before he left the ground. Lovell made peace with the fact that he would have to use the awful NASA fecal bags; for him it was a small price to pay for the privilege of flying in space.

Borman had his own solution to the problem: he simply wouldn't confront it. In his view, if a man couldn't control his own bowels, he couldn't control anything at all, and if that meant controlling them for fourteen straight days, that's what he would do. Through sheer will and orneriness, Borman made it through the first week and then through an eighth day. Partly out of self-interest, even Lovell was beginning to root for him. But no man can hold out forever.

"Jim," Borman said on day nine, "I think this is it."

"Frank, you have only five more days left to go here," Lovell joked. But five days was five too many, and Borman, who believed that any obstacle could be overcome, learned in a new and very primal way that no, not all of them can be.

☆ ☆ ☆

Wally Schirra and Tom Stafford began their own launch day drill on the morning of December 12, which was the ninth day of the Gemini 7 mission and would, if fate and weather and hardware all cooperated, be the first day of the Gemini 6 mission. There would be the usual steak and eggs and juice and coffee for breakfast, followed by the usual suit-up procedure—in proper pressure suits with proper hard-shell helmets—as well as the usual walk to the van and the waves to the press and then the quiet ride out to the launchpad. Walter Cronkite would be in his makeshift newsroom, outdoor reporters would be talking into cameras with the rocket and the gantry in the distance behind them, and the usual tens of thousands of campers and gawkers would be crowding the coast. This time, however, there would be two other spectators, 187 statute miles overhead.

"How's the launch going with 6?" Borman radioed to Houston with less than an hour to go before liftoff. By then, he knew, Schirra and Stafford would be strapped into their spacecraft.

"Oh, 6 is going really well," said Elliot See, a rookie astronaut who'd taken the capcom console shortly after the crew was deposited in the ship. "They're still in the twenty-five-minute hold."

"They've already had a twenty-five-minute hold?" Borman asked, mistakenly believing that an unplanned delay had already dragged on for twenty-five minutes as pad controllers sought to figure out a problem.

"It's a built-in hold," See reassured him. In fact, Mission Control had instituted a launch delay at the T-minus-twenty-five mark to allow for a series of system checks. "Houston said everything's progressing normally."

"Roger," said Borman, relieved.

After more than a week in space, Borman and Lovell were justifiably bored, and the novelty of having visitors had been much on their minds, to say nothing of the challenging flying the rendezvous would involve. And as it happened, the path of their flight would carry them over Cape Kennedy at the precise moment Gemini 6's engines were supposed to light. If the weather held, they would see the Titan's controlled firestorm; from 187 miles up, it would look like little more than the flare of a match head, but they would know their friends were on the way.

Sitting in his television studio, Walter Cronkite was equally eager for liftoff. "This ought to be our most exciting day in space, perhaps exceeded only by our very first space flights," Cronkite told his viewers.

The countdown ticked off smoothly. Perched at the top of the stack, Schirra and Stafford listened as the count approached zero and the glug-glug-glugging began. At precisely 9:54 a.m., they heard the unmistakable roar and shake of the engines igniting.

"My clock has started!" Stafford called over the din, referring to the mission clock on the instrument panel, which was programmed to begin recording every second of the flight the instant the spacecraft lifted off the pad and a tail plug popped loose.

But Elliot See cut in immediately: "Shut down, Gemini 6."

Sure enough, Schirra and Stafford heard the roaring engines go quiet.

Schirra knew the shutdown protocol: without hesitating, he was supposed to yank the D-ring handle between his legs at the front of his seat. A Titan that had lifted even an inch off the ground and then lost its thrust would fall back with a thump that could easily ignite the fuel in the tanks, producing a lethal fireball. Pulling the D-ring would trigger the spacecraft's ejection seats—the most powerful ones ever built—blasting the astronauts out of the capsule at a speed that could outrun the fireball, but only at the cost of a stunning ejection force of twenty g's. While that load might well kill the men, taking that chance was better than getting caught in the fireball, which would definitely kill them.

Schirra had drilled and drilled for this emergency in simulators, but when the moment came, he decided to do exactly nothing. The engines had lit, the clock had started, yet something was missing, something Schirra expected to feel quite literally in the seat of his pants—the kick and the bump of a powerful machine that had begun to move. It was a subtle thing, all but indiscernible against the rumbling, shaking Titan; still, Schirra was certain he hadn't felt it. The cockpit clock was wrong; it had started somehow, but the rocket hadn't moved. The seat of Schirra's pants was right.

The commander stayed calm, and as he looked at the instrument panel, he saw a pressure gauge indicating that the rocket was draining its tanks and returning to its safe mode, just as it was supposed to do after any harmless shutdown.

"Fuel pressure is lowering," he radioed Mission Control, his voice uninflected.

"Roger," See radioed back.

Schirra might have gambled with his and Stafford's lives, but it was a well-considered gamble, and he had won. In the process, he had also saved the last chance for the rendezvous mission. Had he pulled the D-ring, he and Stafford might or might not have survived, but their spacecraft would have had two large holes in it where the hatches had been blown away. There would no repairing the spacecraft or replacing it with a fresh one before the end of Gemini 7's mission.

Overhead, Borman and Lovell had not been able to see the brief fire that had lit and died on pad 19. Elliot See radioed the development up to their spacecraft.

"They got an ignition and a hold kill right afterward," See said, using the crisp, technical phrasing that the moment called for.

Lovell and Borman looked at each other bleakly. They were still alone in low-Earth orbit, and still awaiting the promised visitors.

"Roger," Lovell responded. "This is 7, your friendly target vehicle, standing by."

☆ ☆ ☆

President Lyndon Johnson was already in a foul temper when the news broke that a Gemini 6 mission had flopped yet again. The morning

papers held nothing but bad news, with the *New York Times* reporting
that congressional Republicans and five Republican governors had
issued a unanimous declaration warning that Johnson's escalating conflict in Vietnam was starting to look like "an endless Korean-style jungle
war." This was precisely the comparison Johnson had been seeking to
avoid, and yet now it was in growing vogue. He could take that kind of
defeatist talk from his own party's pacifist left, but given that the GOP
always seemed readier for a scrap with the Communists, this dart carried more sting.

Johnson was just as irritated to learn that an influential group of
thirty black lawmakers and activists in the Democratic Party had put out
a statement demanding a greater voice in the party's policies and candidate selection. Johnson knew that their complaint was justified, but he
had already spent a career's worth of political capital pushing the Civil
Rights Act and the Voting Rights Act through a recalcitrant Congress.
That advocacy had cost him the loyalty of white southern voters, who
had begun stomping off to the Republican Party, where Johnson's equal
rights initiatives were not in much favor. If Johnson couldn't replace that
lost constituency with newly enfranchised black voters, he would be a
pariah in his own party.

And now the space program, which usually offered consolation,
was falling on its face. James Webb and the rest of space agency offi-
cialdom knew that Johnson expected a problem like Gemini 6's
repeated failure to launch to be sorted out quickly. But in case that message hadn't gotten through, the president took the additional step of
issuing a carefully worded public statement. Twice in two sentences he
spoke of his disappointment in this latest debacle. Johnson chose his
words carefully: he made it clear that he did not expect to be disappointed again.

He wasn't. Liftoff for Gemini 6 would come just two days after the
failed launch, thanks to a sharp-eyed engineer who had been assigned
to scour the Titan's innards for whatever had scrubbed the ignition.
Many hours later, he came running back from the launchpad at Cape
Kennedy to report that he had found something deep within the ten-
story stack of hardware—a nickel-sized plastic dust cover blocking what
was supposed to be an open check valve. The cover had been placed

there months before in the assembly plant in Baltimore, when a technician had removed a gas generator for cleaning and, as protocol demanded,
covered the resulting opening to keep out any contamination. But the
cover was easy to overlook, and he'd failed to observe the second part of
that protocol, which called for taking it off it before reinstalling the generator and certifying the Titan as fit to fly. When the engines tried to
light, that little scrap of nothing grounded the giant rocket.

Now the dust cover had been removed, the rocket had been refu-
eled, and the astronauts had repeated their launch-morning drill. On
December 14 at 10:28 a.m., Gemini 6 at last left Earth.

"The clock has started," Schirra shouted as the Titan's engines lit
and, this time, stayed lit. "It's a real one!"

"Trajectory is real good," said See, who was once again manning the
capcom console in Houston.

"Roger, she looks like a dream," Schirra said.

"You're go from here, Gemini 6."

"You got a big fat go from us," Schirra answered. "It looks great!"

Borman and Lovell, once again passing directly over coastal Florida, initially saw nothing, as overcast skies completely obscured the
Cape. But when Schirra and Stafford punched through the cloud cover,
Borman spotted the white thread of the condensation trail following the
rocket and the firefly light of the Titan itself.

"I got it, I got the contrail!" Borman called.

Craning his neck to peek through Borman's window, Lovell spotted
it, too. "It's going to be getting crowded up here," he said.

It would take four orbits for that two-spacecraft crowd to form. And
for nearly six hours, the Earth-based radar tracking that had been minding only Gemini 7 would have to mind 6 as well. Meanwhile, engineers
in Mission Control would be working along with the astronauts and the
computers to pinpoint two tiny machines in the vastness of the Earth's
orbital space and then draw them closer and closer together.

Borman and Lovell donned their hated suits, leaving only the hooded
helmets unzipped and open. NASA rules would never allow two spacecraft to come even remotely within collision distance without ensuring
that the astronauts would be protected if a crack-up occurred and a hull
was breached.

For several days, Gemini 7 had been flying through its orbits in a slow tumble as a means of conserving fuel for planned and unplanned maneuvers. A spacecraft in orbit is not like a jet in flight, which will usually follow its nose wherever it is pointed. As long as Gemini 7's velocity and altitude remained in the right balance, physics guaranteed a stable orbit no matter which way the spacecraft pitched or yawed or rolled. Now, however, Borman took hold of his thrusters and stabilized his ship in a proper, prow-first orientation. It was the only safe way to conduct a rendezvous; almost as important, at least to a pilot, it was the only respectable way.

Gemini 6, with its full tank of thruster fuel and its plan to spend only a little more than twenty-four hours in space, would do the real work in the orbital pas de deux. As Schirra, whose spacecraft was in a slightly lower orbit than Borman and Lovell's, fired his aft thrusters, his speed would increase and his altitude would climb, until his ship had reached the assigned meeting point. While 6 went hunting and 7 awaited its visitor, the radar told both ships that they were indeed drawing close, but so far neither could see the other.

When the two craft were still more than sixty miles apart, with Gemini 6 in the shadow of the Earth's night and Gemini 7 in the full light of day, Schirra turned off the cockpit lights, the better to see the target he was chasing. It was Stafford who spotted what they were looking for first, as the bright sunlight reflected off the back half of Gemini 7, which, unlike the dark front half, was painted a brilliant white.

"Hey, I think I got it," he called over an open mike so that Houston could hear him, too. "That's 7, Wally."

"Negative," Houston responded, relying on the radar readings flickering on a screen rather than the human eyeballs out in space.

"Yes," Schirra replied, now seeing the pinprick of light as well.

"It's either Sirius or 7," Stafford allowed, conceding that he might only be seeing the brightest star visible in the sky at that moment.

But it wasn't a star. Schirra tweaked his thrusters—*blipping* them, as Kraft and the men in Mission Control called it. He vented tiny breaths from the hydrazine jets in the rear of the ship to coax the spacecraft forward, then used counterthrusts from the nose jets when he got going too fast. The ships closed to a handful of miles, then a few thousand yards, and then a few hundred.

Soon the four men were close enough to see each other through the semicircular windows. While Schirra and Stafford were still crisp and clean-shaven after having showered just that morning, Borman and Lovell were uncombed and bearded.

"Hello there!" Schirra called out, beaming. Then he directed his attention to the ground. "We're in formation with 7. Everything is great here!"

Borman smiled back but chose to stick to business. Orbital rendezvous might be a critical step in the long march to the moon, but it was also an exceedingly tricky one and could go badly awry at any second. "I'm reading about ten degrees, one hundred and ten degrees," he said, updating Houston on the orientation of his ship on two of its three axes.

Schirra was in no mood to be serious yet. "We're all sitting up here playing bridge together," he said.

Knowing the rendezvous would last for only a few orbits, however, Schirra soon got to work. There was a lot to accomplish, including the inspection of each ship by the crew of the other. No American astronaut had ever seen his spacecraft from outside while in flight, save Ed White during his brief space walk from Gemini 4, and he'd spent most of the excursion concentrating on keeping himself upright. Now NASA would get a chance to eyeball its machines as they orbited, and the astronauts had been instructed to look for anomalies like gapped welds or seams that might not show in the telemetry but could be disastrous all the same. Inspecting the ships after they returned was much less reliable, because there was no way of knowing if damage had occurred during reentry.

Schirra was surprised to see a tangle of cords and cables streaming from the back of Gemini 7, the remains of the electrical network that had connected it to the Titan before explosive bolts blew the two machines apart.

"You guys are really showing a lot of droop on those wires hanging there," Schirra radioed.

"You have one, too," Borman answered, making it clear on the air-to-ground loop that he wasn't the only commander in the sky today whose ship was not quite parade-ready. "It really belted around there when you were firing your thrusters."

Small American flags and the words UNITED STATES were painted on all Gemini spacecraft, but when earlier Geminis had returned from orbit, much of the flag and the lettering had been burned away. No one had ever been sure whether the fire of liftoff or the fire of reentry had scoured them so badly.

"The flag and the letters are visible," Lovell said now, inspecting 6 as it hovered nearby. "Looks like they're seared as much at launch as they are when you come back."

"Your blue field is practically burned off," Stafford said.

For more than three orbits, the two ships kept their stations— approaching, retreating, flying circles around each other. The remark-able performance offered reassuring proof that the choreography needed for a trip to the moon was indeed possible. In Mission Control, cigars were lit and small flags were waved. Typically, this sort of celebration was reserved for the very end of a mission, but Kraft permitted it to take place early this time.

Schirra, too, had something special planned.

It had escaped no one's notice that there were three Annapolis men in space today, with Borman the lone West Pointer—something the other three had needled him about repeatedly. Now, as Gemini 6 com-pleted another pirouette around Gemini 7's aft end, it reappeared around the front, and this time, its right-hand window was covered with a sign. Schirra had smuggled aboard a piece of blue cardboard bearing bright white letters that read, BEAT ARMY.

Borman dropped his head into his hands and laughed, then made as if to squint at the sign.

"Beat Navy," he said aloud, reading the thing any way he damn well pleased.

Finally, on the first and only day of the mission of Gemini 6 and the eleventh day in the mission of Gemini 7, the station keeping that had been maintained for hours broke off. Schirra backed away, opened the distance he had worked so hard to close, and began easing down to a lower orbit, preparing to reenter Earth's atmosphere.

When the two ships were no longer in sight of each other, Schirra radioed a final transmission—an urgent one, from the sound of it.

"Gemini 7, this is Gemini 6. We have an object, looks like a satellite going from north to south, probably in a polar orbit," Schirra reported. "He has a very low trajectory and a very high climbing ratio. Looks like he might be going to reenter soon. Stand by, just let me try to pick up that thing."

A moment later, crackling across the radio in both Gemini 7 and Mission Control, just nine days before Christmas 1965, there came a tinny chorus of "Jingle Bells." It was performed live, on a harmonica and small set of bells, contraband that Schirra had carried aboard his ship along with his sign.

After he was done, Schirra said proudly, "That was live, 7, not tape."

And then, all business once more, he peeled off and prepared to bring his spacecraft home.

"Really good job, Frank and Jim," he said. "We'll see you on the beach."

☆ ☆ ☆

Less than an hour later, Gemini 6 splashed down in the North Atlantic and was recovered by the aircraft carrier USS *Wasp*. Three days after that, Gemini 7 followed. Its fuel was gone, its power was flickering, and trash filled the narrow space behind the seats, the only stowage area the spacecraft had. Borman and Lovell—unsteady, exhausted, both sorely in need of a very long shower and a very long sleep—waved and smiled gamely as they arrived on the deck of the carrier.

The two astronauts were no better off than their utterly spent spacecraft, but two grueling weeks had come and had gone, the men had survived, and the rendezvous had been achieved. And Borman, who had been making a quiet bet with himself, was delighted to learn that when all the course plotting was done, Gemini 7 had splashed down closer to the *Wasp* than Schirra and his Gemini 6. That would be the end of "Beat Army."

FIVE

GUS GRISSOM SNEAKED FORBIDDEN FOOD INTO A SPACECRAFT twice in his life as an astronaut. The first time was a joke; the second time was a portent.

The time it was a joke, Grissom himself didn't actually do the smuggling. It was his copilot, John Young, who was guilty of that small crime. But Grissom was the commander of the ship, so an offense committed by his subordinate was an offense committed by the superior officer. Either way, the contraband on that occasion was a corned beef sandwich, which Wally Schirra had bought two days earlier at Wolfie's Restaurant and Sandwich Shop in Cocoa Beach and given to Young to take with him on the flight of Gemini 3 in March 1965. It was a stunt very much in keeping with Schirra's sharp sense of humor: half silly joke and half pointed commentary; the jab it delivered was aimed at the terrible pre-packaged meals NASA gave its astronauts to take aloft. The food had been bad enough during the brief flights of the Mercury program; during the long-duration missions of Gemini, the awful meals would be that much harder to bear.

"Where did that come from?" Grissom asked early in the second orbit, after Young reached into his space suit pocket and produced the unsightly, half-flattened sandwich.

"I brought it with me," Young answered. "Let's see how it tastes. Smells, doesn't it?"

The sandwich did smell bad. And although it didn't taste bad, it did release a small starburst of floating crumbs the minute Grissom bit into it, so he wrapped it back up and put it away. When the astronauts came home, the NASA contingent that expected mild misbehavior from its astronauts—which was most of NASA—had a good laugh over the prank. The contingent that fretted over even the tiniest deviation from protocol scowled, reminding all the astronauts that a mere bread crumb could snag in a switch or foul a filter and thus cause a cascade of problems that could lead to disaster.

The second time Grissom sneaked forbidden food inside a cockpit was January 22, 1967. That time the spacecraft was actually a simulator, a training model of the new, three-man Apollo command module on the floor of the North American Aviation factory in Downey, California, where the Apollos were getting built. And that time, the food was a lemon.

Grissom and his crewmates Ed White and Roger Chaffee had been spending a lot of time in both the simulator and in their actual Apollo— which was known on the NASA manifest as Spacecraft 204 but would soon be renamed Apollo 1—in preparation for their February 21, 1967, liftoff. It would be the first flight for the Apollo, and by all appearances, the astronauts would be flying a sweet ship—the most robust, most capable spacecraft NASA and its contractors had ever built. Those appearances, however, were entirely misleading.

To the pilots, the Apollo felt like a slapdash machine. It was temperamental, error-prone, and impossible to work with for more than a little while before something broke down. Then a practice session would have to be halted while technicians crawled inside to fool with a faulty communications system or a dead instrument panel or a life-support component that might fail harmlessly on the ground but would kill the crew if it behaved the same way in space. Repairs were made as needed, but they were patchwork affairs—workarounds and fixes made on top of earlier fixes, rather than the harder undertaking of ripping out the offending systems, redesigning them, and reinstalling them only when they worked right. The Mercury and Gemini spacecraft had had their design problems, but those had been glitches in ships that from the beginning had looked and felt sound. The Apollos—perhaps because of their complexity or perhaps because of the rush to get astronauts on the moon before 1970—never earned that confidence.

A disgusted Grissom would complain to the technicians, and then he'd complain to the technicians' bosses, and then he'd complain to the NASA bosses. They would all confer among themselves and promise Gus that they would fix the problem, and still the Apollo simulator— which worked only as well as the Apollo spacecraft itself—got no better. So Grissom decided to make his point a different way. After another long day of trying to get the practice ship to work the way it was supposed to, he climbed out of the ship and left a lemon perched atop it before walking silently away.

That Gus, the engineers said to one another with indulgent smiles. *Always a little prickly.*

But Grissom had a right to be a lot more than prickly. After the joint flight of Geminis 6 and 7, the space program had been moving along at just the clip NASA had planned. Between the December day when Borman and Lovell splashed down in the North Atlantic and the following November, five more Geminis had flown, one every other month, finishing up with Lovell returning to space as the commander of Gemini 12. Lovell had been joined by Buzz Aldrin, a promising rookie from the third astronaut class, and the mission had been a confidence-boosting finale for Gemini. Now all of the program's most difficult goals—space walk, rendezvous, docking—had been performed both confidently and competently.

Well before that final spacecraft in the Gemini line returned to Earth, manufacturers had begun cutting metal for the Apollo line. Both the Mercurys and the Geminis had been built by McDonnell Aircraft in St. Louis, and NASA liked the people there just fine. They knew what the space agency needed, delivered what had been ordered, and understood that although they might own the factory and employ the workers, NASA was both the customer and the boss.

But McDonnell couldn't come along for Apollo. For one thing, the company had its hands full. McDonnell's people had been working at a dead sprint since 1960, and they did not have the resources to keep the Gemini assembly line running and build the Apollos at the same time. According to NASA, only three months were supposed to elapse between the late-1966 launch of Lovell's Gemini 12 and the February 1967 launch of Grissom's Apollo 1. It was an almost impossible job, and only a company that was not already overworked could accomplish it successfully.

Even if McDonnell could have handled the assignment, giving another project to the same manufacturer wouldn't have been right. The space program—like road construction or dam building or rural electrification—was a public works project, and though a private developer might be free to contract out work to the same builder again and again, NASA had Congress to answer to, and Congress had hundreds of constituencies to answer to. There were plenty of other manufacturers in cities all around the country, and they would all want to know why St. Louis kept finding NASA's favor again and again.

So this time the work went to North American Aviation. When the contract was signed, that seemed like a perfectly suitable choice, but when the work started coming in, it had turned into something else entirely. For starters, there were the folks in what the company called the human factors division.

Test pilots thought of a man and his plane as two discrete things. One was a deeply dumb machine that had been built to do what it was told; the other was a brilliant aerialist whose job it was to tell the machine what to do. But North American didn't see things that way. Man and machine were simply two parts of the same system, and the engineers in the Downey offices believed it was their responsibility to make sure that they performed together properly.

Straight off, that presented problems, particularly for the astronauts themselves. The first pilot to bump up against North American's new— and unwelcome—way of doing business was Frank Borman. After his and Lovell's Gemini 7 marathon, the two men went in different directions. Lovell preferred to stay in the Gemini rotation until he could get a command of his own, an ambition Borman completely understood. After all, no good pilot is happy to serve at the pleasure of another pilot when he figures he could do the job just as well himself. For his part, Borman had had his chance in the left-hand seat, and after fourteen days in a Gemini cockpit he had no interest in spending still more there. Instead, he went off to work in the Apollo program, serving on-site in Downey as pilot liaison and thus helping to shape the ship he and the other astronauts would be flying next.

One day early in his time there, Borman hopped into the Apollo simulator to give it a trial spin. He was exceedingly displeased by what he found.

Trying out the hand controller that fired the simulated thrusters, he realized that the engineers had designed it upside down. When he pulled back on the handle, the Apollo's nose pointed down; when he pushed forward on it, the ship pointed up. That was exactly the opposite of the intuitive way airplanes work—and the way the Gemini spacecraft did, too.

He summoned an engineer and pointed with dissatisfaction at the offending handle. "You've got the polarity on this thing reversed," he said, speaking as politely as he could, since the unsuspecting man would surely be embarrassed. "It goes down when you pull up and up when you push down."

"Oh no, that's the way we're going to fly it," the engineer responded brightly. "It makes rendezvousing easier. When you pull back on the stick, the nose goes down but the target will seem to go up. This way it'll be like you're flying the target, not the spacecraft."

Borman was speechless. First there was the man's presumptuous use of the word "we"—as if there were any question about exactly who would be flying the spacecraft. More important, there was his apparent ignorance about how the astronauts who would be sitting in the cockpit did their work.

"You've got pilots who have been flying jets," he said, just managing to contain himself, "and that's not the way they do it."

"That's the way our human factors people say we should do it," the man replied, seeming unmoved.

Now Borman was genuinely angry. "Well, that may be the way *you're* going to do it, sitting on your ass as an engineer," he snapped. "But that's not the way *we're* going to do it."

Borman was doing precisely the job he'd been assigned to do: stopping nonsense before it found its way from the woolly brain of an engineer to the metal innards of a spacecraft. In his next call to Houston, he reported how the factory fellows wanted to design the thrusters. Half an hour after the call, the handle was fixed.

☆ ☆ ☆

A mere simulator, however, was not remotely the biggest problem in the Apollo factory. Everywhere in the production process, it seemed, rules were being flouted and speed was taking priority over safety. A lot of the North American engineers had learned their craft in the so-called black programs—the classified programs—of the military, which mostly involved building unmanned vehicles, particularly satellites and missiles. It was hard work, performed to fine tolerances, but none of it had entailed designing systems that would have to keep a man alive. Their missiles, in particular, didn't even have to work terribly long or terribly well; all they had to do was fly from silo to target and blow up when they were supposed to.

Word of quality-control errors in the Apollo spacecraft came back to NASA headquarters with disturbing regularity, and the matter eventually found its way up to Chris Kraft, who wanted to get his own quiet proxy on the floor at Downey. Kraft decided that what he needed was a Brain Buster, someone from the old Hampton days who could take a machine apart with his eyes and put it back together in his head in such a way that the redesigned product would almost always be better than the original. So he called on John Bailey, his former boss at NACA, who was about the finest engineer he had ever met. He asked if Bailey might be willing to take a look at the North American operations and write a memo telling him what he thought.

Bailey did, and his report was bleak, going on in great detail about each system and subsystem that was a cause for worry. But it was his from-the-gut assessment of the operation in general that was most worrisome to Kraft.

"This hardware is not very good," he reported. "The cabling is being stepped on when they work on the spacecraft. There's no protection for it. The people are not very good at checking this thing out. They're not very good at trying to maintain some semblance of the fact that a human being is going to be in this machine. I'm telling you, it's not good."

But NASA was racing against John Kennedy's fast-closing deadline and a cranky Congress that was growing increasingly reluctant to fund a moon program at the same time it was pouring money into the widening war in Vietnam. The choice for the space agency seemed to be fly now or wait for perfect hardware and fly never. So despite Bailey's memo, the work went on and the spacecraft got built. To Kraft and others, their only answer was to pick a crew that inspired complete confidence, one that should be able to handle whatever their troubled spacecraft might throw at them.

Grissom was one of only a handful of men who had been in space twice. He had piloted both the Mercury and Gemini, and both times he had flown an early iteration of the spacecraft and helped wring out its problems. White had flown once, on Gemini 4 in 1965, when he made his historic space walk and displayed a remarkable steeliness throughout. The walk looked like grand fun, with pictures showing the white-suited astronaut silhouetted against the blue of Earth and the black of space. But the twenty-three minutes outside the ship had, in fact, been brutal, a constant struggle against zero-g physics that made maneuvering vastly more difficult than it looked. Even getting safely back inside had proved harrowing: White's hatch had refused to close for a full five minutes. In the end, he'd had to rely on sheer muscle to yank it shut, exerting himself so much that he fogged his visor completely.

Chaffee, the rookie, might not have traveled in space yet, but he knew how to pilot a flying machine and how to stay alive in one even when all manner of forces were conspiring to kill him. In 1962, when the United States and the Soviet Union had come within an eye blink of nuclear war over the sudden appearance of ballistic missiles in Cuba, Chaffee had been one of the Navy pilots who'd flown reconnaissance missions over the missile sites. Getting shot down or even chased off could easily have become the casus belli for war, but he had kept his head and flown his missions and helped the Americans win that stare-down with the Soviets.

As a crew, the three astronauts tried to appear sanguine about their upcoming space mission, but they had no illusions about the jalopy of a ship NASA was handing them. As their launch date drew near, the space agency began circulating publicity photos of the three men who would take America's moon ship on its first test run in Earth orbit. In one series

of pictures, the men posed in pilots' jumpsuits while sitting at a table
with a model of the Apollo spacecraft in front of them, smiling in coun-
terfeit confidence at the camera.

In a photograph that was never circulated, however, Grissom, White,
and Chaffee showed their true feelings about their spacecraft: they posed
with their heads bowed and hands held together in prayer. Then, to
make sure their message got delivered to the right people, they inscribed
it to Harrison "Stormy" Storms, the North American engineer who was
overseeing the Apollo project. Like the rest of the top people at North
American, Storms was well aware that astronauts on the factory floor
were regularly making calls to NASA to report yet another problem in
the Apollo spacecraft.

"Stormy," the inscription read, "this time we're not calling Houston!"

☆ ☆ ☆

On January 27, 1967—five days after Grissom made the statement with
his lemon and less than a month before the scheduled launch of Apollo
1—NASA planned to conduct what was known as a plugs-out test. The
exercise would begin when a fully suited crew climbed inside the spacecraft, which was already on the pad and mounted on top of its Saturn
1B booster. With the spacecraft operating on its own internal power system, the crew and the controllers would perform a dress rehearsal of the
launch sequence.

Two more steps would ensure as much launch day authenticity as
possible. The first involved the Apollo's atmosphere, which was made up
of 100 percent oxygen, as it would be in orbit, instead of the roughly
22 percent oxygen, 78 percent nitrogen mix of Earth's atmosphere.
Humans need only the oxygen to stay alive, so the designers had decided not to outfit the spacecraft with tanks of inert nitrogen, since that
would merely add weight.

They knew that in the vacuum of space, the oxygen in the cockpit
would be pumped to a pressure of slightly less than five pounds per
square inch—just a third of the nearly fifteen pounds per square inch
found at sea level, but really all an astronaut needed. On the launchpad,
however, the interior pressure would have to be much higher, to prevent

the force of the outside air from squeezing and damaging the hull of the low-pressure spacecraft. So for the plugs-out test, the Apollo was inflated all the way up to 16.8 pounds per square inch. If anyone was concerned about the fact that fire loves oxygen—especially pure, high-pressure oxygen—that concern did not cause NASA to halt the test.

The second authentic condition involved the hatch, which, once the astronauts were on their backs and in their side-by-side seats, would be behind them and directly over White's head. In the event of an emergency, Grissom, White, and Chaffee would be best served by a hatch they could open in a hurry. That would allow them to tumble out of the ship, onto the floor of the white room, the little work space at the end of a swing arm at the top of the gantry tower. The white room surrounded the Apollo when the ship was on the pad and swung back out of the way before liftoff. An easy-open hatch, however, would not be suitable for a cockpit with so much internal pressure. For that, engineers designed a double hatch—an inner one and an outer one—and sealed it with multiple latches. If a pad emergency occurred, the man in the center seat would open the latches with a ratchet and then detach the inner hatch, pull it in, and lay it down on the floor. Only then could he open the outer hatch. The commander, in the left-hand seat, would assist if needed. The Apollo 1 astronauts had practiced this sequence many times, and no matter how hard or efficiently they worked, it took time.

The night before the plugs-out test, Wally Schirra, who was Grissom's backup for this mission, went out to the launchpad with Grissom and spent some time inside the spacecraft running a few final tests. When he climbed out, he shook his head.

"I don't know," he said to Grissom. "There's nothing I can point to, but something about this ship just doesn't ring right."

It was a damning judgment for a pilot to deliver, suggesting a vehicle that didn't have discrete, fixable problems but instead had sweeping systemic ones.

And then Schirra added a warning: "If you have any problems, I'd get out."

Grissom promised he would.

The plugs-out test at last began at 2:50 p.m., after the crew had settled into their seats and the double hatch had been closed and sealed. The exercise ran slowly and haltingly. The day's most nettlesome problem was one that had occurred in earlier tests, too: balky communications. White and Chaffee could make out the transmissions coming through their headsets, but only through a storm of static and a lot of dropped words. Grissom's communications line, for reasons the engineers couldn't manage to discover, was even worse.

Before the exercise began, Deke Slayton had offered to climb into the ship along with the crew and spend the entirety of the test period in the lower equipment bay—the small work space beneath the foot of the couches, the astronauts' cot-like seats—to see if he could solve the communications problem. But the plugs-out test was supposed to be as authentic as possible, and since there wouldn't be four men jammed into the three-man ship on launch day, there wouldn't be four men today, either. Instead, Slayton stayed in the launch control center at the Cape, listening as best he could to the garbled communications coming down from the ship.

At 6:20 p.m., an exhausted crew and the exhausted ground teams working at both Cape Kennedy, in Florida, and Mission Control, in Houston, got a short break as the simulated countdown went into what was supposed to be a ten-minute hold while the communications breakdown and other glitches were addressed. Kraft, who had been shuttling back and forth between Mission Control and his nearby office as the countdown started and stopped throughout the afternoon, was now back at his console at the rear of the control room, listening to both the chatter on the ground and the transmissions from the spacecraft.

"How are we going to get to the moon if we can't even talk between three buildings?" Grissom groused at just a few seconds shy of 6:30 p.m. It was one of the rare moments his voice got through clearly.

"They can't hear a thing you're sayin'," White said, his tone bemused.

One minute and fourteen seconds later, the people on the ground did hear something the crew was saying. What they heard was Chaffee shouting, "Hey!"

Next they heard White screaming: "Fire! We've got a fire in the cockpit!"

Then they heard Chaffee shouting: "We have a bad fire!"

And finally they heard Chaffee again, this time screaming: "We're burning up!"

In the white room, the technicians could see frenzied motion through the spacecraft's windows. They also saw the flickering light of what was unmistakably a fire.

"Get them out of there!" yelled Donald Babbitt, the pad leader and the chief of the white room. The men around him leapt toward the spacecraft and began wrestling frantically with the hatch. Intense heat radiating from the metal hull forced them to turn their faces to avoid the full power of it.

As the noise and frenzy in the white room increased, the transmission from the spacecraft became ominously still. Now the voices in the headsets of the men in the control rooms were filled only with the words of the launch controllers themselves—shouting questions to one another at the Cape, staying silent in Houston, where they could do nothing.

"Crew egress!" called Chuck Gay, the lead test conductor in the launch, following the book and telling the crew to do what they could not possibly do, which was to get out of a spacecraft that had just become a furnace.

"Blow the hatch!" shouted Gary Propst, a communications technician nearby. "Why don't they blow the hatch?"

In the white room one more voice rang out, though no one ever figured out whose it was. "Clear the level!" the person called. He was using the agreed-upon language to order all of the people outside the ship to back away—or, if possible, to run away—because the spacecraft was in danger of exploding.

Moments later, the Apollo gave off a sound and a blast of air like a bomb, showering the white room with flaming debris and setting fire to loose papers on clipboards and desks. Inside the spacecraft, the fire rushed toward the freedom of the opening, completely engulfing the astronauts. There would be no surviving that; the crew was lost.

Kraft heard every word the dying men said. Slayton—who was sitting at a console, not folded up in the equipment bay of the burning ship—heard it all, too, as did every other man in the two control centers. But no one would ever be able to agree on exactly what the astronauts' final words were. Even when recordings of the last few seconds of the men's lives were played, different people remembered hearing different things that the tapes didn't capture, which was entirely possible, given the unreliability of the communication systems.

What many of the men insisted they heard, even though the tapes didn't catch it, was one of the astronauts—a professional pilot to the last, a man who knew that as long as you are able to communicate with your flight controllers, you must keep them apprised of the condition of your ship—saying as levelly as the circumstances would allow, "I'm reporting a bad fire." The report would be duly noted.

NASA announced the deaths of the men within the hour, and the networks preempted Friday evening programming with regular reports on the accident at the Cape. The space agency reassured the public that Grissom, White, and Chaffee had died almost instantly; though only a small consolation, that was still a mercy. But the official story wasn't true: the three men had lived for a long time as these things go—at least twenty-one seconds, judging by the biometric readings and the activity in the ship that the technicians in the white room had seen through the windows, as well as the data recorded by the spacecraft's motion detectors.

For twenty-one seconds, the astronauts knew what was happening to them and fought to save themselves. To anyone at NASA who had been paying attention throughout the year, the immolation of Grissom, White, and Chaffee was not an accident but an inevitability—equal parts tragedy and disgrace.

☆ ☆ ☆

Gene Kranz, one of NASA's top flight directors, was not in Mission Control on the night of January 27. Moments before the fire struck, he was home getting ready for dinner—a dinner that, he had been told, would involve grape leaves. That prospect did not especially please him, but there wasn't much he could do about it. He and his wife, Marta, had just had their third child, but throughout her pregnancy, Kranz had not been able to do much to change the seven-day-a-week schedule he'd been working while trying to get his team ready for the February launch of Apollo 1. On the morning of January 27, he left for the Cape early, as he always did, and spent much of the day helping to prepare for the plugs-out test. But since the exercise didn't necessarily require his presence and he knew it could easily run twelve hours or more, he had promised Marta that he'd come home in time to change and take her out for dinner at the restaurant of her choosing.

Recently, a Greek place had opened in the formerly gritty Houston Ship Channel. The ship channel still functioned as a huge working port, but it had also begun to attract a number of new businesses, most of which served the young families that had been flowing into town ever since the space program had gotten rolling. This new restaurant's main attraction was a selection of entrees wrapped in grape leaves, and if Kranz could not understand the cuisine's appeal—why not just eat the food inside the leaves with a knife and fork and include the greens in a salad if you must?—Marta saw things differently. Either way, he knew he owed her the night out.

Given the hours he'd kept most of his career, Kranz reckoned he owed Marta a lot of nights out. A former Air Force pilot, he had flown patrol operations during the Korean War and served as a flight test engineer at Holloman Air Force Base, in New Mexico. He'd gone straight from there to NASA in 1960 and had been working at a frenzied pace for most of the seven years since. It was Marta herself who had encouraged Kranz to answer the ad in *Aviation Week* that said the government was looking for personnel for a "space task group." Once on board at NASA, Kranz had climbed fast, and had been groomed as a flight controller by Chris Kraft himself.

It was Kraft who'd taught Kranz how a mission ought to be run and how it ought not to be run, and he'd taught it the best way possible: on the job. Kranz had done splendidly in the Mission Control rank and file, one of many men working one of many consoles. But he'd outgrown the job quickly, and before long he'd been bumped up to flight director, the man at the center console, in whose hands the success of the flight lay for each of his eight-hour shifts.

One of Kranz's first missions in that rarefied position was Gemini 5, an eight-day flight in August 1965 that held the space endurance record until Gemini 7 shattered it later that year. Early in the mission, the

Gemini 5 astronauts began reporting that their spacecraft was having a number of problems; the cryogenic tanks that fed super-cold oxygen and hydrogen to its fuel cells were causing particular headaches. The pressure in the tanks was continually reading lower than it should, and if the tanks failed, the main power system would go with them.

During one particularly harrowing stretch, Kranz arrived in Mission Control for his shift. Kraft, who had been the flight director on the previous shift, handed the command over to him with little more than a hello and a good-bye, then prepared to leave the control room for some much-needed rest. Before turning the center console over to the next shift's director, Kraft would ordinarily talk his replacement through the options for the eight hours ahead, particularly when a shaky ship like Gemini 5 was flying. But this time he simply shucked his headset, stood up, and turned to walk away.

"Chris," Kranz called after him. "What do you want me to do?"

Kraft turned back with an expression that showed—to Kranz's eyes, at least—more than a little irritation. "You're the flight director," he said sharply. "It's your shift. You make up your mind."

After handling that challenge and others with aplomb, Kranz had become one of Kraft's favorites. Along with controllers Glynn Lunney and Milt Windler, Kranz was one of the three most recognized workhorses at the main console in Mission Control.

On the night of the Apollo 1 plugs-out test, however, Kranz had left the space center early, gone home, and begun changing for the dinner he'd promised Marta. Before he was done, he heard a pounding at the front door. Expecting the babysitter, he frowned at the excessive noise. Still half-dressed, he hurried downstairs and opened the door to see his neighbor Jim Hannigan, a deputy director of the space center's lunar module division.

"There's been an accident at the Cape," Hannigan said without preamble. "I heard it on the radio. They think the crew is dead."

Kranz tore upstairs, got back into his work clothes, and told Marta what he knew. Then he hurried out to his car and raced to the campus of NASA's Manned Spacecraft Center. When he arrived and ran to the Mission Control building, however, he found that it had already gone into lockdown, a well-practiced procedure that was designed to minimize outside distractions during a crisis. He tried all of the doors that would normally take him into the building; they were all locked. He lifted the security phone to summon a guard; although the line was live and rang as it should, no one answered.

Finally, Kranz ran around to the back of the building and at last found a guard near a freight elevator. Kranz talked fast, flashed his badge, and made it clear who he was and where he needed to be—which was inside the building, not outside on the lawn. The guard saw reason and took him up. When an out-of-breath Kranz at last entered Mission Control, the scene there chilled him.

At the center console was Kraft, talking in low tones to the flight surgeon at the Cape. Near him was John Hodge, chief of the flight control division. They both looked grim, but nowhere near as much as the pale and shaken younger men at the other consoles. Hodge had worked in test flight programs and had long ago learned what death looked like. So had Kranz, who had also flown in combat, and Kraft had been at NASA long enough to have prepared well for disaster.

The other men in the room had had no such training, no such toughening. They were boys, really, straight from engineering schools; their average age was just twenty-six. In the classroom way that fresh PhDs know things, they'd understood that they were participants in a game with mortal stakes. But book learning and bloodletting were two different things, and tonight the space program was bleeding badly.

Kranz and the other senior figures in the room put the junior men back to work, instructing them to secure their consoles so that every switch and dial remained in exactly the position it was in when the fire struck. Somewhere in those thousands of settings might be critical information that would be needed in the investigation to follow—and there would surely be an investigation.

Not long after that bleak forensic business was finished, the controllers decamped as one. With unspoken agreement, they drove to the Singing Wheel, the bar near the space center favored by NASA personnel. Many of the older hands also called the place the Red Barn, because a well-loved watering hole deserved not only its proper name but a nickname, too, and to its regulars from NASA, this place had a red barn feel about it.

As always, Lyle—few of the controllers knew his last name—was working behind the bar. He knew what had happened at the space center, and he took care to shoo his other patrons out as soon as the flight controllers arrived. Once they had the place to themselves, they did what they had come here to do, which was to get exceedingly, sorrowfully drunk. Lyle let them stay for as long as they needed to, closing up only after the last of them had staggered home to grieve alone.

☆ ☆ ☆

The weekend that followed the fire was nearly unbearable, not least because most of the Mission Control team had nothing to do until Monday except think about what had happened at the Cape on Friday. But when Monday morning finally arrived, the controllers—red-eyed or not, sleepless or not—were expected back at work. When they got there, Kranz was ready for them. He and Hodge called a meeting in the main auditorium in Building 30 on the Manned Spacecraft Center grounds, and attendance was mandatory.

After the controllers took their seats, Hodge spoke first. He talked about what had been learned over the past couple of days; thus far, only a little was known about either the fire's source or the extensive under-lying problems in the spacecraft. He announced the names of the men who had initially been assigned to the investigation team, and he talked a bit about how much time would be needed until flights could resume. He conceded that he had no idea if the long-standing deadline for a lunar landing was realistic anymore, but he assured his audience that the NASA administrators would make every effort to come as close as possi-ble to landing on the moon by 1970. Then he turned the stage over to his colleague.

Kranz had spent his weekend thinking hard and getting angry, and he wasn't about to waste an opportunity to drive home a lesson that had come at a tragically high cost. Too often in the previous months, he told the silent controllers, potential problems had been dismissed with a casual "that can't happen" wave. Maybe the ship had a balky breaker, but it would never cause a fuel cell to fail in flight. Maybe those new pyrotechnics were a little temperamental, but they could never make a parachute fail to deploy. And as for pumping pure oxygen into the cockpit, it had never caused any problems before, had it?

But what if it did? What would you do then? That was the critical question no one had been raising. It was not good enough to ask what you would accept. Instead, you had to ask what action you would take today to prevent the failure from ever happening. The answer you gave should always satisfy one final question: What is the very best thing to do in this situation?

Kranz made it very clear to the men in the Houston auditorium that the best thing to do was not what had been done in the months leading up to the deaths on Friday. Not a person in the room—himself included—had been tough enough, competent enough, or accountable enough. Every one of them had seen one or more of the problems the Apollo spacecraft had been having; everyone had heard the stories about the wreck that was rolling off the lines at North American Aviation. And not one of them had stepped forward and spoken up.

"We had the opportunity to call it off," Kranz said sharply. "We had the opportunity to say, 'This isn't right. Let's shut it down.' And none of us did."

With enough smarts and enough skill, Kranz told them, no mission ever had a reason to fail. It might have problems; it might not achieve every one of its goals. But failure, in all its abject awfulness, cannot be on the menu of possibilities.

"From this day forward," he said, "we will stand for doing everything right, literally being perfect and competent."

Kranz turned to the blackboard behind him and wrote the words "tough and competent." Then he turned back to his young charges.

"I want every one of you to go back to your offices and write those words on the top of your own blackboards," he ordered. "You are not to erase them until we've put a man on the moon."

With that, he set down his chalk, turned on his heel, and left the stage. The lesson was over.

☆ ☆ ☆

The news of the death of the three astronauts came to Frank Borman the same way it had come to Gene Kranz: with a knock at the door. Borman, Susan, and their boys were staying at a friend's cabin on a lake in Huntsville, Texas, and they were just starting into a long weekend that would at last give Borman a break from the Apollo steeplechase. The race to get the ship off the pad less than three months after the last Gemini spacecraft flew had been exhausting, and the prelaunch stretch to come promised only to be worse.

The knock on the cabin door that Friday evening was not remotely expected. Borman had mentioned to someone or other where he and the family would be spending the weekend, but he didn't rightly remember who. He had said it in passing, more as idle conversation than as a means of informing the space agency about where they could find him if the need arose.

He opened the door to find a Texas Ranger or a highway patrolman—he couldn't tell exactly which.

"Colonel Borman?" the officer asked.

"That's me," Borman said.

"I have word from the space center. You're to call Mr. Slayton immediately."

Borman thanked the man and hurried to the phone. He knew even before Slayton answered that the news could not possibly be good, but what the chief astronaut told him was far worse than he had imagined. Borman closed his eyes and felt his stomach drop out; for a moment he could say nothing at all. He would grieve for Grissom and Chaffee, but he would bleed for his friend Ed White. Borman did not give his friendship easily, and now he barely knew what to do with the loss.

He found his voice. "How did it happen, Deke? What went wrong?"

"We don't know," Slayton said. "But you're on the team that's going to investigate this thing. Be at the Cape tomorrow morning."

Borman agreed, hung up the phone, and quietly told Susan and the boys what Slayton had just told him. They packed hurriedly and drove back to Houston; without even stopping at their own home, they pulled up across the street at the Whites' house, finding a space to park wherever they could among the astronauts' sports cars and government sedans that were already crowding the curb. They went inside—no need to knock today, just as there was never a need to knock when an astronaut was in space and the family's home was filled with well-wishers around the clock, serving the cake and sandwiches and coffee during the day and the casserole and potato salad and whiskey in the evenings. On those bright, busy days, everyone would be walking the wire between excitement and terror, a balancing act that would last until splashdown.

But tonight terror had won. The feared thing had happened, and the people in the room showed it: they were drawn and hollow-eyed, subdued and deeply sorrowful. Susan rushed to Pat, who was surrounded by the wives, and they embraced like sisters. Borman joined the astronaut cluster, exchanging grim nods and murmured condolences. The government people, in their own small group, spoke in the low and purposeful tones they used when working at something important. And what they were working on right now were the funerals. That, from what Borman could hear, was also what Pat was talking about with Susan— when she could work her words through her hitching breath.

The decision had already been made that all three astronauts would be buried at Arlington National Cemetery sometime in the coming week, just as John Kennedy had been buried there a little more than three years before. It was precisely the kind of honor the men deserved, except that Ed White would have wanted nothing to do with Arlington, Kennedy or not. Like his father, White had been a West Point man, and more than once he had made it clear to Pat that since he had learned to be a soldier and taken his commission at the academy, he wanted to be interred there if he lost his life in the race for space. Washington, however, had decided there would be just one funeral for the entire crew; it was neater that way, especially since President Johnson planned to attend.

Susan summoned Borman and told him of the problem. Pat pleaded with him to set things right. He nodded.

"Ed will be buried at West Point," he promised Pat.

"But they said there could only be one service," she responded.

"There will be two," he said.

Then he walked to the government men, asked for the phone number of whichever protocol office had dispatched them here, and dialed it impatiently.

"This is Frank Borman," he said when the man answered. "I'm at Ed White's house. It is his family's wishes that he be buried at West Point."

The man began to protest, explaining that the Arlington arrangements were already being made.

"I don't care," Borman said. "Ed will be buried at West Point, as the family wants. Now go make *those* arrangements, because that is what is going to happen."

He hung up the phone, leaving it rattling in its receiver. Borman would be at the Cape in the morning as ordered, but he would pack his Air Force dress blues as well. Later in the week, he would be traveling to West Point to help carry his friend to his grave.

☆ ☆ ☆

Not long after Borman arrived at the Cape, he visited the launchpad and climbed inside the ruined Apollo. The spacecraft was still perched atop its rocket to nowhere, and from a distance, in the right light, the bright white ghost ship still looked ready to fly someplace grand.

By now, the bodies had been removed and sent off to the medical examiner's office. Yet there wouldn't be much of an examination: it was already clear that the men had died not of burns but of asphyxiation. The fire had flashed through too fast to have consumed much of the cloth of their suits, but the smoke and the fumes from a thousand melting materials would have been impossible to survive for long.

Grissom had been found partly in his left-hand couch and partly slumped toward White's in the center, suggesting that when he died he was performing his part in the evacuation drill by leaning over to assist White as he tried to open the latches on the hatch. White died in his own couch doing that work, with one arm dropped across his face, as if he were trying to protect himself from the poisonous smoke that had filled the ship. Chaffee, too, was still in his seat; his assignment would have been to maintain radio contact with the ground throughout the escape, the likely reason his voice was heard first.

By the time Borman climbed inside, plastic drop cloths had been hung over the instrument panel and spread on the seats, keeping the death scene pristine until the plastic could be rolled back bit by bit during the mechanical necropsy to come. It would be an exceedingly painstaking job.

The position of all of the switches on the instrument panel had been preserved and noted, just as they had been on the consoles in Mission Control. Since the state of any one wire could have birthed the spark that in turn started the fire, each one—that is, of those that hadn't been burned or melted away—would be traced through the spacecraft and to its source. And since a bolt that had been insufficiently tightened could have caused something to shake loose, and a bolt that had been tightened too much could have crimped a connection, the torque needed to unscrew each one would be recorded to the inch-ounce.

The deconstruction would go on for several months. Ultimately, the work would reveal the exact sequence of small, avoidable problems that had led inexorably to the tragedy on the launchpad.

At precisely 6:31 p.m., a spark jumped from a wire on the far left of the spacecraft, beneath Gus Grissom's seat. The wire ran beneath a little storage compartment with a metal door that had been opened and closed many times without anybody noticing that each time the door moved, it abraded a little more of the wire's insulation, finally leaving the exposed copper free to spark at will. When the wire did spark during the plugs-out test, it ignited a small fire that stayed small for only a second or two. Accelerated by the pure oxygen, the flames climbed along the left-hand wall of the spacecraft, feeding on fabric and netting used for storage. That wall was the worst possible place for a fire to erupt, since it prevented Grissom from reaching a latch that would have opened a valve and vented the high-pressure atmosphere, thus slowing the flames.

Unimpeded, the fire proceeded to consume anything around it that would burn—the paper of the flight plans, the cloth of the seats, the Velcro and plastic and rubber that were everywhere. It fed on the space suits of the men themselves, spreading over Grissom and toward White, who by now was struggling with the hopeless ratchet and the locked-tight hatch. As the temperature climbed inside the ship, the pressure gauges—whose readings were being recorded by computer—rose from 16.8 pounds per square inch to 18 and then 20; seconds later, they soared to an explosive 30. At that point, the spacecraft exceeded its structural limits, causing a weak spot on the floor on the far right of the cabin to do what the person who'd shouted, "Clear the level!" had feared, which was to rupture.

The explosion ripped apart not only the doomed spacecraft but all the plans NASA had so carefully drawn for its steady march to the moon. The astronauts would be mourned by the nation, the space agency would be pilloried in the press, and both houses of Congress would investigate the accident and issue scalding reports. To no one's surprise, NASA indefinitely suspended all planned flights of the Apollo spacecraft.

☆ ☆ ☆

Only when the deconstruction of the accident was complete did Borman move on to the real job he had in the investigation, which was to go out to Downey to work on the factory floor. Appointed by administrator Jim Webb and Deke Slayton, Borman was the astronauts' chief representative and advocate. It was far too late in the game to sack North American Aviation and start over with a new contractor. Instead, the place would have to be reformed.

Borman was not the only astronaut there; to his mind, in fact, there were too many of them milling about. John Young, who had flown with Grissom on Gemini 3 and with Mike Collins on Gemini 10, prowled the floor as if he were the company's CEO, looking over shoulders, checking on production lines, and firing off memos to anyone who would read them, demanding this or that change and insisting that it happen right now. Some got read, some didn't; privately, Borman started calling them Young-grams, and he treated them with a mixture of respect, appreciation, and occasional exasperation.

Wally Schirra was on-site too, but in a sense he wasn't there at all—not the real Wally, at least. As the commander of Apollo 1's backup crew, Schirra would have been sitting in the left-hand seat if Grissom had happened to have a head cold on the day of the plugs-out test. And now that the prime crew was gone, it would be his team, which included rookies Donn Eisele and Walter Cunningham, who would be the first to fly the Apollo. Schirra figured that he had a bigger stake than anyone else in making sure the spacecraft got fixed, which meant that the jolly Wally everyone knew was nowhere to be seen.

"You roasted three men in that thing already," he would snap at anyone on the floor at Downey who objected to interference from an astronaut. "You're not going to roast me."

Ultimately, the combination of grief, guilt, and stress became too much. Before long, the breakdowns began—not in the hardware this time but among the men doing the work. Scott Crossfield, the legendary test pilot who had been the first human being to reach twice the speed of sound and had been the only man in the air who could give Chuck Yeager a run, had come to work for North American before the fire as part of the Apollo quality-assurance team. Crossfield had never had a terribly sure hold on his whiskey, and now the stuff got hold of him instead. As the investigation went on, he became angry, argumentative, and ineffectual. The end finally came the day a North American team wanted to roll out a second stage for the Saturn rocket that Crossfield didn't think was ready.

"If they try to take that thing out of here, I'll lay down in front of it," he barked.

All work within earshot of Crossfield stopped. Heads turned his way, but he held his wobbly ground.

"What should I do?" a North American director asked Borman. Crossfield might have worked for the company, but the company—especially now—was answerable to NASA. And sloppy on drink or not, Crossfield was still Crossfield; a mere engineer could not reprimand a legend.

"Have someone fire him," Borman answered.

"I can't do that," the man answered. "He's . . . Scott Crossfield."

"I don't care who he is," Borman said. "Have him fired."

The man nodded warily and passed the directive on. Later that day, Scott Crossfield—pilot, hero, almost a Yeager—lost his job.

One of the civilian contractors came apart as well, succumbing to the pressure of too much work, too little sleep, and who knew how much grief and guilt. Attending a meeting in a North American office one day, he began—to the horror of the other attendees—drawing an organizational chart of heaven, taking care to include a position he labeled "Big Daddy." An ambulance was summoned, and the man was collected and carried away.

Through it all, however, the Apollo spacecraft was slowly being redesigned and rebuilt. The wiring was rewoven; the hatch was replaced with a new model that could be opened by one person in seconds; the flammable Velcro was eliminated. Every shred of ordinary paper was replaced by fireproof paper, and every scrap of ordinary cloth was replaced by fireproof beta cloth. Combustible coolant was replaced, soldered joints were reworked, vibrational tests were improved, and the 100 percent oxygen atmosphere used on the pad was replaced by a 60-40 nitrogen-oxygen mix. Every single step in the quality-control and checkout processes was also rethought and redrafted.

Most important, the engineers and the human factors people at North American would never again be allowed to slip the leash of the space agency.

"Nobody's going to put anything in the spacecraft unless NASA management approves it," said Borman—who, as far as North American was concerned, *was* NASA management.

And nobody ever did.

SIX

1967–68

Nobody fully appreciated what a bright, hopeful charm the moon had been until it was suddenly moved back out of reach. The steady climb to a lunar landing had been bracing, with the spaceships flying and the calendar cooperating, and for a while it seemed almost certain that NASA would meet the challenge President Kennedy had thrown down. Now there was nothing; the deadline was less than three years away, and the ships, for the moment at least, were grounded. Worse, what remained was a national mess that stretched well beyond NASA.

If 1967 had been birthed in a fast, hot fire at the Cape, other, larger fires soon seemed to be breaking out everywhere else. Lyndon Johnson could point proudly to his Voting Rights Act and his Civil Rights Act, but laws counted for little if the people they were meant to benefit were still suffering in rural poverty in the South and in broken, stultifying ghettos in the North. That summer, as the nation endured a record-setting heat wave, more than 150 U.S. cities erupted in race riots, and many of them were put to the torch. New York burned in late July, when cars were overturned and set ablaze in the largely segregated neighborhood of Harlem. Newark's fires raged for six terrible days; riots killed 26 people and injured a thousand. Detroit followed, with 43 deaths, 1,189 injuries, and some 7,200 arrests. Firefighters abandoned a hundred square blocks of the city, which quickly burned to the ground, the flames hurried along by winds of twenty-five miles per hour that struck at just the wrong time.

In October, fires of a different sort were lit, as more than one hundred thousand antiwar protesters, fed up with the endless combat in Southeast Asia, marched on Washington, rallying first at the Lincoln Memorial and then across the Potomac River at the Pentagon. Their loathing for Johnson and his Vietnam adventure was palpable, and the signs they carried were savage. Some read, LBJ—HOW MANY KIDS DID YOU KILL TODAY? and, much worse, WHERE'S OSWALD WHEN YOU NEED HIM? Kennedy's assassin, like Kennedy himself, might be long dead, but the blood in which Johnson's presidency had begun seemed to be rising back up and threatening to drown him entirely.

Within NASA, the mood was equally bleak. Everyone understood that 1967 was going to be a washout, at least in terms of getting American astronauts off the launchpad. In an agency known for its bravado, doubt crept in. How could it not? The moon project depended on a spacecraft that had proven itself a literal deathtrap, a Saturn V rocket that had never carried men into orbit, and a lunar excursion module that hadn't even been completely built.

Against the background of these delays, many at NASA were nevertheless impatient to see the Apollo program start flying again, no one more so than the astronauts themselves, Frank Borman among them.

It had been Borman's own choice to cash out of the Gemini program early, but he had done so in order to get a head start on Apollo. Yet as 1968 began, he was entering his third year without a mission after his sole trip into space aboard Gemini 7. Jim Lovell, John Young, Pete Conrad, and Tom Stafford had two notches on their guns, all of them flying once in the right-hand seat and then going back up, this time in the left. The Gemini program had ended with Lovell, Borman's former junior, commanding Gemini 12, and that voyage had proved to be an all but flawless finale for the two-man spacecraft. Within the exceedingly ambitious corps of astronauts, the two-time fliers were increasingly starting to look like the real stars of the incoming Apollo class. Borman and the other one-timers somehow seemed second-tier.

Worse for Borman, word had started going around NASA that Bob Gilruth—one of the space agency's founding wise men and the director of the Manned Spacecraft Center in Houston—wanted to take him out of the cockpit altogether. Not long after the fire, it had become clear that Joe Shea, chief of the Apollo program almost from its inception, had to go. It wasn't entirely Shea's fault that the early development of the Apollo spacecraft had been such a sloppy business. Still, most people agreed that if the day-to-day operation of the program had been the responsibility of someone like Chris Kraft or Gene Kranz, heads would have been knocked together earlier and the shoddy work exposed a lot sooner. Maybe that wouldn't have saved Grissom, White, and Chaffee, but the men would at least have had a better chance of surviving.

In April, Shea was moved to a desk job at NASA headquarters in Washington. In July—recognizing the posting as the banishment it was—he resigned from the agency altogether. He was replaced on an interim basis by Gilruth's deputy, George Low, but a permanent replacement was needed fast. Gilruth, known as a good judge of talent and a master at getting men to take the jobs he needed them to do, couldn't say exactly what he was looking for in a new Apollo program chief, though he would know it when he saw it—and during the recovery from the fire, he saw it in Borman.

Gilruth had a high opinion of the work Borman had done on the factory floor at Downey, and soon it became an open secret inside NASA that he hoped the astronaut would abandon his dream of traveling to the moon and instead take over the program that would send other men into space instead. But Borman had other ideas: he had been grounded once before, because of a bad eardrum, and he was not about to let the same thing happen again, this time because of a job well done. As soon as his work at Downey allowed, he returned to the training regimen for Apollo and immersed himself in the curriculum, as much to catch up with what he had missed as to make clear to anyone who was watching that he had come to NASA to fly a spacecraft, not a desk.

Finally Gilruth called Borman in for a meeting. He did not waste time before getting to his point: "You know, Frank, I'd very much like to have you run the Apollo program."

Borman looked back at Gilruth respectfully but said nothing, hoping his silence would say everything.

Perhaps it did. Gilruth immediately seemed to sense where the conversation was going; as he always did in situations like this one, he set about making sure that all parties came away feeling that the right decision had been made. So after a pause he said, "But I also know you want to keep flying, and you don't have the administrative experience we need right now."

Borman nodded solemnly in response, trying his best to convey sincere disappointment at this very bad news.

Moving on quickly, Gilruth told him that he had thought a lot about the position and decided that Low might be up to taking it on permanently. "But if you're really interested," he said, "I'll reconsider."

"No," Borman replied, perhaps a little too quickly, "I couldn't agree with you more."

The two men shook hands on the decision, and with that, Borman was officially back in the flight rotation.

☆ ☆ ☆

In early 1968, a full year after the fire, the preliminary mission assignments for the first several Apollo flights were finally released. The space agency's astronauts had waited for this moment with the nervous excitement of junior high school students anticipating their homeroom assignments, but when Borman saw the slot he'd been given, he was crestfallen. Despite his high standing, despite his yeoman work on the factory floor at Downey, despite the fact that the head of the Manned Spacecraft Center had wanted to put the entire Apollo program under his command, he had once again been handed a stinker.

The first mission was still going to be Wally Schirra, Donn Eisele, and Walt Cunningham's maiden flight of the Apollo spacecraft in Earth orbit. Nobody knew what the exact numeral of any Apollo flight would be yet—a lot depended on how many unmanned tests of the Saturn V rocket spacecraft were run—but it was looking like Schirra's crew would be known as Apollo 7.

After that would come Gemini veterans Jim McDivitt and Dave

Scott and rookie Rusty Schweickart—part of the third astronaut class—aboard Apollo 8, test-driving both the Apollo spacecraft and the lunar excursion module (LEM), once again in Earth orbit. Even for men who were eager to peel off from the home planet and fly to the moon, Apollo 8 was a sweet assignment, primarily because getting a first crack at the LEM would be part of the deal.

The lunar module, by even the most charitable definition, was a madman's machine, different from every aircraft or manned spacecraft ever built before it. Even vehicles designed to blast off from Earth and soar into space still had to fly *through* the atmosphere before they got there. That meant they needed a certain aesthetic elegance, with the smooth-skinned, tapered-nose backswept shape that was necessary to provide lift and keep air resistance to a minimum. But the LEM didn't care about any of that. Tucked inside the uppermost stage of the Saturn V booster, it would be carried to space and wouldn't see starlight until it was out of Earth orbit and on its way to the moon, when it would be extracted from the rocket by the Apollo spacecraft. That meant it could be designed strictly for the job it had to do, which was flying through the vacuum of space, landing on the moon, and then getting off again.

The result was the opposite of elegant: the LEM was a twenty-three-foot-tall, four-legged insectlike beast. Its triangular windows looked like nothing so much as a pair of angry eyes, and its trapezoidal mouth served as the door through which the astronauts would crawl to reach the ladder that would take them down to the lunar surface. The ship was covered in crinkly, reflective insulation, a material that also served as the walls of the crew's cabin—walls that were no thicker than about three layers of aluminum foil.

For the LEM, weight was everything, since even the Saturn V could lift only so much. From the beginning, the designers at the Grumman Aircraft plant in Bethpage, Long Island, fought to give the machine as much capability as possible, while shaving off every available ounce.

Part of that job was easy, because the LEM was actually a two-part ship. Its first task would be to land on the moon with its four legs and a powerful descent engine. When it was time to take off, the bottom half of the ship would serve as a launch platform, with explosive bolts and a guillotine system cutting the cables and other links to the top half, allowing an ascent engine to carry the remains of the spacecraft—which was essentially the crew cabin—up to lunar orbit.

That cabin would be as spartan as possible. Seats would weigh too much, so they would simply be dispensed with. Besides, the LEM would be doing its work only in the weightlessness of space or in the moon's one-sixth gravity, so the crew could stand. That also made it possible to eliminate the wraparound windows the initial designers had envisioned; though they would have maximized the astronauts' field of vision, they'd be very heavy. If the men were standing, they could press their noses— or the faceplates of their helmets—right up against the little triangular portholes and get the same effect.

The wires used in the ship were the narrowest gauge possible, so that they made their own small contribution to weight reduction. But they were also exceedingly fragile and broke like spider's silk if the builders didn't handle them just so. To reduce the LEM's weight even further, the engineers gave all of the metal surfaces a round of what they called chem milling—chemically burning away half a millimeter of thickness here and a quarter millimeter there. This painstaking work made almost no difference in the weight of any one spot, but over the total body of the spacecraft, the reductions added up.

The Apollo 8 crew would have the job of taking this mass of foil origami on its first flight. McDivitt and Schweickart would climb inside, separate from the Apollo spacecraft, and kite around for a few orbits, while Scott piloted the command module alone. If, for some reason, the LEM could not rejoin the Apollo, Scott might be the only one coming home. Unlike the command module, which was equipped with a sturdy heat shield, a LEM would go up like a piece of flash paper if it tried to reenter the atmosphere. People outside of NASA might dismiss the entire flight exercise as just another jaunt around the Earth, but the astronauts knew the complex piloting it involved, and Scott took to describing Apollo 8 as "the connoisseur's mission."

Next up would be Borman's shot. Once again he would get the commander's left-hand seat, which pleased him. The affable Mike Collins would be in the center seat, and off to the right would be Bill Anders, an eager rookie from the third astronaut class. Borman didn't know Anders terribly well, but the newcomer already had a reputation as a lunar

module wunderkind, someone who had made it his business to understand the strange machine better even than the men who had designed
it. If McDivitt's crew did their job well, Borman's mission ought to be
able to fly to the moon. But to Borman's great frustration, what the mission planners had in mind for Apollo 9 was little more than a repeat of
Apollo 8.

It would be the same basic drill for more or less the same number of
orbits, with the one exception that Apollo 9 would end its mission by
climbing to an altitude of 4,600 statute miles and practicing the high-dive
reentry that astronauts would have to execute when they returned from
the moon. That wasn't just beanbag. A Gemini spacecraft orbited the
Earth at 17,500 miles per hour and then eased back into the atmosphere
by tapping its brakes until it was no longer moving fast enough to stay aloft.
But an Apollo spacecraft returning from the moon would slam into the
atmosphere at nearly 25,000 miles per hour, aiming for a narrow keyhole
in the sky just a couple of degrees wide. Too steep an approach would
cause the ship to burn up on reentry; too shallow and it would skip off the
atmosphere like a rock off a pond and then go careering into space forever.

Borman didn't see why McDivitt's crew couldn't just tack the high-
altitude reentry onto the end of their flight. Wouldn't the three Apollo 8
"connoisseurs" want that little finale to their time in space? The first
flight of the LEM was historic; the second would be understudy stuff. In
the wake of the fire, however, the cautious fellows in the flight planning
office weren't about to take chances again. The journey to the moon
would be a slow, incremental process. Not until Apollo 10, 11, 12, or later
would anyone go anywhere near the lunar neighborhood.

For Lovell, meanwhile, the Apollo prospects looked brighter. He
would have a small role on Borman's Apollo 9, serving as backup to
Collins in case the prime crewman had to be scratched from the lineup.
But the odds of that happening were slim: Collins kept exceedingly fit and
did not do a lot of the hard drinking and fast driving some of the other
astronauts did. If he was assigned to fly a mission, there was very little
chance he wouldn't go.

That meant that Lovell would get a starting spot on a later, distinctly
better, flight: the center seat on Apollo 11, between Neil Armstrong on
the left and Buzz Aldrin on the right. Apollo 11 might or might not be a

lunar landing, and even if it was, Lovell would have to wait overhead in lunar orbit while Armstrong and Aldrin flew down in the LEM and got their boots covered with moon dirt. Still, 11 could easily be a bona fide moon shot, an adventure that looked a lot better than the Earth-orbit dog-paddling Borman had been assigned. Perhaps hanging around Gemini a little longer had been the smart play after all.

☆ ☆ ☆

The men at Grumman might do brilliant work when designing and building the lunar module, but no one was going anywhere if the Saturn V rocket, which was being built by multiple other contractors, didn't get off the ground—and that was going to take some doing.

Traveling from the launchpad in Florida to the plains of the moon was, at bottom, a business of using a very big rocket to get a very large payload moving very, very fast—faster than human beings had ever traveled before. That called for a rocket unlike any that had ever been built, and NASA never tired of wowing the world with the size of the monster they had invented. The Saturn V stood 363 feet tall—the height of a thirty-six-story building, or sixty feet taller than the Statue of Liberty, or longer than a football field, including the end zones.

The monster's weight was just as impressive. Fully fueled, a Saturn V tipped the scales at 6.5 million pounds, or a third again as heavy as a Navy destroyer. A Navy destroyer, however, moved on its belly, sliding across the surface of the ocean, a prisoner of gravity. The Saturn V flew.

Each of the giant rocket's five first-stage engines gulped three tons of kerosene and liquid oxygen fuel per second, burning through more than five hundred tons in the 168 seconds of life given to that stage before it was done with its job. At 41 miles of altitude, the first stage was jettisoned, allowing the second to take over. That stage carried five smaller engines that burned through 340,000 gallons of fuel in just 384 seconds. After that came a third stage with a single engine that would give the astronauts the final kick into Earth orbit and would later be lit a second time to sling them out toward the moon.

This extraordinary machine was also extraordinarily dangerous. When the first stage was lit and the rocket was just beginning to haul itself off the pad, the engines would be generating 160 million horsepower—as much energy as would be produced if every river and stream in the United States were running through a single hydroelectric turbine at once. Only a nuclear explosion could produce a louder man-made sound than a Saturn V taking off. And if the rocket went rogue and exploded on takeoff, well, the physicists had pretty good numbers on that, too: they calculated that the blast would create a fireball 1,408 feet in diameter, burn for 33.9 seconds, and generate temperatures in excess of 2,500 degrees.

The first launch of the monster machine—the mission called Apollo 4—had taken place on the morning of November 9, 1967. The show was sensational. Spectators crowded the Florida coast the way they usually would only for a manned flight, and they got what they came for.

"My golly, our building's shaking here! Our building's shaking!" shouted Walter Cronkite in his temporary studio, three miles from the launch site. Cronkite had always reacted to rockets the way a small boy reacts to diesel locomotives, and he didn't care who knew it. Inside the makeshift newsroom, the picture window facing the launchpad—built extra large to allow the newsmen to see what they were describing—began rattling in its frame. The crew in the booth leapt forward to hold it in place. Cronkite, clearly having the time of his life, helped them.

"The roar is terrific!" he called as the building's ominous-sounding tremors became audible to viewers. "This big glass window is shaking, we're holding it with our hands! Look at that rocket go! Part of our roof is coming in here!"

Things weren't much calmer in the firing room that was Cape Kennedy's Mission Control. Plaster dust rained down from the ceiling, settling on consoles that were typically kept free of the slightest bit of dirt. The engine's thunder—which reached 135 to 140 decibels, exceeding the pain threshold, and produced tectonic vibrations that were detected by seismometers as far north as New York—could easily be heard inside the reinforced building. The controllers whooped and cheered at their dusty desks. Even Wernher von Braun, who was using binoculars to watch the launch through the firing room's more robust window, allowed himself a moment of jubilation.

"Go, baby, go!" shouted the German designer, who had never been heard to call anyone or anything "baby" before.

It took less than twelve minutes for the third stage of the Saturn and the unmanned Apollo spacecraft to reach Earth orbit, and once the rocket and its cargo got there, they did everything that was asked of them. The Apollo beamed healthy vital signs down to Houston, reassuring administrators that if astronauts had been aboard, they'd be just fine. Just under three hours—or two orbits—into the flight, the third stage fired to boost the Apollo from its safe, 119-mile-high orbit to a nosebleed altitude of 11,000 miles; after this, the command module separated and executed the acrobatic, deep-space reentry flawlessly, landing less than ten miles from the prime splashdown site near Hawaii.

"That was the best birthday candle I ever had!" Arthur Rudolph, a von Braun assistant who'd turned sixty that day, exulted to reporters after emerging from the launch room.

That perfect trip, however, would turn out to be the last gift NASA would get from its big rocket for a while. The launch of Apollo 5 came just over two months later. This time a smaller, two-stage Saturn 1B rocket carried a lunar module—given the honorific LEM-1—to orbit so that NASA could see how the little lander worked. In what might have been a portent, the top stage of the Saturn 1B was the same one that would have helped carry Grissom, White, and Chaffee to space and was part of the stack that had been on the pad the night the men died. The stage hadn't been damaged by the fire, and with Congress ready to pounce on any sign of waste, NASA wasn't about to throw the thing away just because it was unlucky.

Even so, the mission did seem to be snakebit. The Saturn 1B did its minimal job, getting the LEM into orbit, but the lander itself underperformed. Its computer misread the commands being sent to it, causing its descent engine to cut off far too early. And when its ascent stage separated, the computer tripped up again, never recording that half the mass of the ship was gone, leaving the ascent stage tumbling through its orbit as it tried to regain its balance in the wrongheaded belief that it weighed twice as much as it actually did.

The newspapers were charitable, declaring the mission a qualified success, if only because the LEM had finally flown. But on a moon mission, the LEM would have almost a zero tolerance for error, and a lander that flew like the LEM-1 would have surely killed its crew.

The last of the unmanned flights—or the last if NASA had any hope that Apollo 7 would finally return American astronauts to space—was set for three months later, on April 4, 1968, when the Saturn V would fly for the second time. The hope was that with Apollo 6, the big booster would give the space agency its swagger again.

This launch of the Saturn rocket was as spectacular as the first. But two minutes into the flight, unstable fuel pressures in the first stage caused the rocket to bounce up and down like a pogo stick as it flew, oscillating so rapidly that it threatened the structural stability of the whole stack. Any astronauts who had been aboard would have been injured, perhaps badly; as it was, the violent shaking knocked loose two of the petal-like panels in the upper compartment that would have been carrying a lunar module if one had been on board.

No sooner had the first stage been jettisoned and the second stage lit than it, too, became unstable. That caused two of the five engines to cut out completely and—according to the alarming telemetry data coming down from the ship—also caused a twelve-inch structural I-beam inside the stage to bend; this, in turn, threatened to shake the engines loose completely. That would lead to what the engineers would call a "catastrophic failure," and what everyone else would call an explosion. Somehow, the second stage continued to function and the third stage limped into orbit with the Apollo spacecraft atop it, but there was no dressing up the outcome of this mission.

"This was a disaster," Kraft said after he emerged from Mission Control, daring anyone inside NASA to try to prettify what had just happened. "I want to emphasize that. It was a disaster."

Seventeen months had now elapsed since an American astronaut had flown in space. Only twenty months remained before one of them was supposed to set foot on the moon. For the moment, America's space program seemed to be moving in reverse.

☆ ☆ ☆

If there was anything good about Apollo 6, it was the mission's brevity: it began at one second after 7:00 a.m. eastern time and ended less than two hours later with the splashdown of the command module in the central Pacific. And although it was a terrible day for Kraft, for NASA, and for the Apollo program as a whole, the press reports on the disappointing mission would prove to be fleeting.

Ten hours after the Saturn V's command module splashed down, at 6:05 p.m. central time, the Reverend Martin Luther King stepped onto the balcony outside of room 306 at the Lorraine Motel in Memphis, a city that was in the midst of a sanitation workers' strike. King was in town to deliver a speech on the workers' behalf. Standing with him on the balcony was Ben Branch, a bandleader and jazz saxophonist who was due to perform at a union gathering that evening.

"Ben, make sure you play 'Take My Hand, Precious Lord' in the meeting tonight," King said. "Play it real pretty."

At that moment, a bullet from a Remington Model 760 rifle—fired from a rooming house some fifty yards away—struck King in the cheek, jaw, neck, and jugular vein. He was rushed to St. Joseph's Hospital, but there would be no recovering from his wounds. He was pronounced dead just over an hour later.

Even before King's assassination, it seemed certain that 1968 was going to be a blood-soaked year. At the end of January—during Tet Nguyen Dan, the Vietnamese New Year—more than 70,000 North Vietnamese and Viet Cong troops flooded across the demilitarized zone, striking thirteen major targets in South Vietnam and one hundred smaller towns and villages. The attacks would lead to 165,000 casualties on both sides and more than 14,000 civilian casualties. The term "Tet Offensive" would immediately enter the vernacular as shorthand for both the hopelessness of America's involvement in Vietnam and the spent force that Lyndon Johnson's once-promising administration had become.

Johnson himself seemed to know it. On the evening of March 31, he had delivered a televised address to the nation, ostensibly to talk about the conduct of the war. But he clearly had more on his mind: at the end of the broadcast he dropped a political bombshell, announcing, "I shall not seek, and I will not accept, the nomination of my party for another term as your president."

If Johnson thought that sacrificing his political life would calm the furies that were raging that year, he was mistaken. In the wake of Martin Luther King's murder, ten major cities—including Baltimore, Chicago, and Washington, D.C.—erupted in riots, resulting in dozens of deaths, thousands of arrests, and tens of millions of dollars in property damage. The fight to succeed Johnson turned ugly, too, as the various wings of his party tore themselves apart. In June, Senator Robert Kennedy, who had become a candidate for president only three months earlier, was assassinated in a Los Angeles hotel, just minutes after winning the California primary. The delegates he won that night were pledged to a ghost, and once again the mourners and the drummers and the catafalque for a Kennedy appeared. And still the violence raged on: in August, at the Democratic National Convention in Chicago, twenty-three thousand club-wielding police officers descended on ten thousand antiwar demonstrators, leading to what would later be called a "police riot."

Through it all—quietly, doggedly—the engineers and planners and astronauts at NASA labored on. In their windowless rooms, on their guarded bases, they lost track nearly altogether of the flaming world around them. To the extent that they noticed the explosions of violence at all, it was only as a dull thumping beyond the walls, a dim flashing through a very thick scrim.

SEVEN

Summer 1968

NOBODY EVER TOOK CREDIT FOR THE GREAT BRAINSTORM THAT changed everything. That made sense, because it's highly unlikely that among the four hundred thousand men and women working in all of the offices, labs, factories, and universities in the sprawling effort to get Americans on the moon before 1970, only one person saw the range of problems on the space program's plate and suddenly realized that there was a brilliant if daft solution to them. Everyone knew about the jumpy Saturn V and the troubled lunar module and the Apollo spacecraft that had yet to carry a human being off the launchpad, and given the rush of the calendar and the impatience of the public, at least a few of the exceedingly creative people working for the space agency must have been struck by the same idea—a wonderfully insane way to set it all right at once, to push the whole pile of hoarded chips across the felt and onto black and spin the wheel and take the chance.

But it was George Low who first gave voice to the brainstorm. He spoke about it in August 1968, three months before Wally Schirra was supposed to take his Apollo 7 spacecraft up to Earth orbit. That July, the first completed LEM arrived at the Cape and was the same kind of mess it had been in the factory: full of defects, still overweight, literally too weak to stand on its own four feet. True, it had been designed to be only strong enough to support its bulk in the one-sixth gravity of the moon, but that helplessness still seemed like the perfect metaphor for so fragile and flawed a ship. No matter how hard the technicians worked, the LEM would clearly not be ready for its November or December shakedown flight with Jim McDivitt's Apollo 8 crew.

Exactly what Low said was remembered one way by Low himself and another by Chris Kraft and a third by Bob Gilruth. But all three men were present the day the chief of the Apollo program asked his two colleagues to listen to what he had been thinking. And what Low said, in effect, was: *You know our schedule? Let's screw our schedule. Let's go to the moon in sixteen weeks. And let's give the job to Apollo 8.*

☆ ☆ ☆

Frank Borman had been gratified to learn that Michael Collins was assigned to fly with him on Apollo 9. There might have been a few astronauts who trained as hard as Collins did, but no one surpassed him. Collins had flown only once before, aboard Gemini 10, and in the year leading up to that launch he had vanished into training like a man obsessed. His performance on the mission itself had been flawless: he had walked in space and helped dock the spacecraft with an Agena rocket, now that the Agena was at last flying true.

With Apollo 9 approaching, Collins was hard at work again, and no one doubted that he would do as well on this mission as he had on the last. But Collins had a secret, one even he didn't know until the day he was playing handball and felt strangely heavy-legged. It was nothing he could quite describe; he just wasn't as quick or limber as usual. He tried to ignore the feeling; after all, he was thirty-seven now, so he was bound to have lost a little spring. That bit of denial worked for a while, but then came the buckling left knee and the numb lower leg and the strange way cold water hurt when it touched his calf, while hot water—even hot enough to scald—produced no sensation at all. And then the feeling began traveling up his leg and into his thigh.

At last, right about the time when George Low was sharing a brainstorm with Bob Gilruth and Chris Kraft, Collins went to see the NASA medic. Collins was a pilot, so it wasn't his way to think too much about the truly awful possibilities the NASA doctor would be considering—multiple sclerosis or muscular dystrophy or that disease that killed Lou Gehrig when he was almost exactly the same age Collins was now. Collins was thinking only about the risk of being found unfit for flight; the way he saw it, nothing could be worse than being stripped of his wings.

Within an hour of his arrival in the infirmary, with X-rays already taken and the medical people already scrutinizing them, Collins and the doctor thus had two completely different reactions to what the film revealed. The doctor was profoundly relieved to discover that the astronaut's problem with his leg was being caused by the bone spur he'd found between Collins's fifth and sixth cervical vertebrae, a problem he could fix with a comparatively simple operation. Collins, meantime, was knocked flat by the hard truth that until he had the surgery and then rehabilitated his back, he would be a ground pounder and nothing more.

It fell to Deke Slayton to give Collins that news officially. Slayton urged him to schedule the surgery as soon as possible so he could return to the flight rotation quickly—something Collins didn't need to be told.

It also fell to Slayton to tell Borman that one of his men had been scratched from the lineup, and then to call Jim Lovell into his office.

"Jim," Slayton said when Lovell arrived, "Mike's got this bone spur and needs to have surgery. I'm switching you to 9 and, assuming Mike's up to it, he'll be taking your spot on 11."

Lovell was familiar with what the protocol dictated he do in a situation like this, which was to say, *Okay, Deke*. So he said, "Okay, Deke," and then he even added a "happy to do it."

Actually, this news didn't make Lovell happy at all. Like Borman, Lovell knew that Apollo 9 was just an Apollo 8 rerun with a high-orbit cliff dive at the end, while Apollo 11 gave him a better shot at going to the moon. Now, having once spent fourteen days going around the Earth on Gemini 7, he would be logging another ten days on Apollo 9, never leaving the planet's neighborhood.

And there was something else he didn't like about the new mission, something he probably could have discussed with the chief astronaut, because an old military flier like Slayton would understand. Until that moment, Apollo 9 had been an all–Air Force crew: Borman, Collins, and the rookie Bill Anders all wore USAF wings. Slayton was Air Force, too. Lovell, meantime, was Navy.

If Borman had been outnumbered three to one during the Gemini 6 and 7 fandango—with that sweet BEAT ARMY sign in Gemini 6's window—Lovell would now be the man in the minority. That was not the kind of thing that would significantly affect the job the astronauts did, but it could potentially turn perfect crew cohesion into something subtly less than perfect. Whigs may dine with Tories, Red Sox may drink with Yankees, but really, a man prefers his own.

Still, Lovell understood that the new assignment had its compensations. Flying sooner was always better than flying later, and as long as the Apollo program ran deep enough into the teens—Apollo 13 or 14 or 15—Lovell should get his shot at the moon anyway. And he certainly wouldn't mind flying with Borman again, even if he did chafe at being back in a subordinate seat after having been in command on Gemini 12. As for Anders, well, Lovell got a kick out of the rookie.

So Lovell resolved to make the most of Apollo 9, just as he had made the most of Gemini 7. That was what you did when you got a flight assignment in the Navy or at NASA—and then you went back to work.

✻ ✻ ✻

Bill Anders, just shy of thirty-five in the summer of 1968, had quite literally been born into the military. His father had been a Navy lieutenant stationed in Hong Kong and living in married officers' quarters when his son was born. Like Lovell, Anders had graduated from the naval academy, but afterward he had gone into the Air Force to get his wings—a decision a pure Annapolis man like Lovell never could quite understand.

Anders, like Borman, had come to NASA by way of Chuck Yeager. Also like Borman, he had not left Yeager on the best of terms. In 1959, the young fighter jock had applied to Edwards Air Force Base to study test piloting, but Yeager had turned him down.

"We're only accepting men with advanced engineering degrees these days," he'd said, waving Anders out of his office.

Anders, who didn't like to be told no, promptly enrolled in the Air Force Institute of Technology at Wright-Patterson Air Force Base, in

Ohio. For three years he studied hard and well. Then, degree in hand, he returned to Edwards and knocked on Yeager's door again.

"That criterion has changed," Yeager said. "Advanced degrees don't count anymore as much as flying time does." Flying time, of course, was exactly what Anders hadn't been accumulating while spending most of the past three years in a classroom.

Still, Yeager agreed to at least consider the overeducated pilot, and Anders held out hope that he would ultimately pass muster. In the meantime, he returned to New Mexico, where he was working at the Air Force Special Weapons Center. Not long after, in the spring of 1963, he was driving his Volkswagen bus in the desert near Albuquerque when a reporter on the radio announced that NASA was accepting applications for its third class of astronauts and that this time around it was changing its requirements. As long as a candidate held an advanced engineering degree, he would no longer have to be a test pilot. Anders pulled over to the side of the road and waited a full fifteen minutes for the news to repeat to make sure he had heard it right. He had.

As soon as he could, Anders wrote away for the proper forms, submitted his application, and was called in to begin the grueling selection process. By now, the doctors had gotten more sophisticated—and more diabolical—than they had been for the first two astronaut classes. Added to the usual tests of stamina and raw intelligence were more behavioral games and complex psychological drills.

There was the pinball-like aptitude machine, for example, which required the astronaut candidates to solve brain teasers or spatial reasoning puzzles. That was difficult enough, but they also had to work on up to ten problems at any one time, meaning they would almost constantly be throwing levers or switches or responding to blinking lights by pushing the proper buttons. Some of the problems were easy, but others demanded the candidate's complete attention. Anders quickly reasoned that since he was being graded not just on accuracy but on speed, there was no mathematical advantage to wasting more than a moment on the hard ones. He decided to simply guess on those and then machine-gun his way through the easier ones. The strategy worked: he blew the doors off the test, winding up with a score that was 150 percent better than the next-highest finisher.

He took a similarly tactical approach to the written tests, relying on what he and his college classmates had called "the mumbling theory." Professors, they'd figured, were as easily bored as other human beings, and few things could numb the mind like grading essays. So Anders would pour himself into answering most of the questions thoroughly and well, and then offer up vague but persuasive gibberish—easy to write but exhausting to read—for the handful of questions that he reasoned would take the most time and yield the worst results. The selection board, he suspected, would give the mumbles at least a serviceable C, while rewarding him with A's on the rest of the questions. By his reasoning, nothing he did counted as breaking the rules. Instead, all of it counted as deftly playing them.

The strategy must have worked, because on October 17, 1963, his thirtieth birthday, he got a call from Deke Slayton congratulating him on his acceptance to the astronaut corps. Anders had dreamed many times in the previous months of getting this call, and he responded to it exactly as he'd rehearsed in his mind—gratefully and respectfully.

Just two days later, he got another call he had fantasized about for the past several months as well. It was from Yeager, at last giving him what the old colonel must have assumed was very bad news.

"I'm sorry, Anders," Yeager said. "You're a fine candidate, but you're just not right for the test pilot school."

Anders—all cocky thirty years of him—knew what he ought to say in this situation and what he ought not say. He chose the impolitic response.

"Well, Colonel," he said, "I appreciate your call, but I have a better offer anyway."

Yeager asked what that offer was, and Anders told him. Wasting little time, the colonel, who did not suffer insolence lightly, put his considerable influence to work by making the necessary phone calls and pressing the necessary decision makers in an effort to get Anders bumped from the astronaut program. He didn't succeed, but word of his campaign got out and put a cold fright into Anders. That may or may not have been what Yeager intended, but it was what the young pilot needed.

Anders took to astronaut training well. He had plenty of time for it, too: with thirteen other astronauts in his class, plus the nine in class two and the four still flying from class one, there wouldn't be many flights to go on anytime soon.

Every new astronaut was supposed to pick a few specialties, and Anders concentrated on lunar geology and radiation shielding. Mostly, though, he decided to make himself an expert on the LEM. Partly that was pure ambition, but partly it was because he absolutely loved the bizarre little machine. In some ways, the LEM was put together just like he was: it broke not a single rule of flight design, but it played them all creatively, daringly, even whimsically. Anders could think of no finer thing than being able to fly one of the landers someday—even if he had to take it for a spin in Earth orbit first.

By that point, the young astronaut had already learned one critical lesson: if you could master the LEM, the only way you'd lose a chance to land on the moon was if you somehow screwed up. And after having come close to sabotaging his career with his wisecrack to Yeager, Anders was not about to repeat the mistake.

☆ ☆ ☆

George Low had a lot of reasons for taking his Mad Hatter chance and proposing an early flight to the moon. Some of those reasons he could talk about, and some he couldn't. The steady march of the calendar and the problems with the LEM were no secret. Nor was the grumbling coming from a budget-conscious Congress.

Then there was the problem almost no one knew about, one that involved the Soviet Union. The death of the Apollo 1 crew had shaken both countries' space programs badly. The night of the fire, Soviet ambassador Anatoly Dobrynin—who loved all things related to space travel, whether the ship carried a hammer and sickle or the Stars and Stripes on its side—happened to be attending a reception at the White House. Dobrynin was there to celebrate the signing of what was popularly known as the Outer Space Treaty, though its formal name was the Treaty on Principles Governing the Activities of States in the Exploration and Use of Outer Space, Including the Moon and Other Celestial Bodies. The United States, the Soviet Union, and the United Kingdom were the first to sign, but the pact was open to any other country, provided they were willing to abide by its rules. The treaty's most important covenants required the signatories to agree not to militarize space and to offer all necessary assistance to any astronauts who landed on their territory and repatriate them and their spacecraft promptly.

The White House reception was a must-attend affair, although even if it hadn't been, Dobrynin would have shown up, if only to meet all the American astronauts who would be present. Just as the reception was ending, word of the disaster at the Cape got out, and the Soviets, with their sharp ears and their sprawling intelligence web, heard the news even before it was widely announced. Dobrynin wrote to Lyndon Johnson the next day and expressed his genuine sorrow.

But sympathy for the loss of the Apollo 1 crew did not mean that Moscow was blind to an opportunity. A slowdown in the American program gave the Soviets a clear lane, at least for a while. And by the summer of 1968, they were making the most of it. Word pinging across America's own intelligence web had it that Russian space engineers had been aggressively testing their new manned spacecraft, the Zond, with plans to launch the next two ships, Zond 5 and Zond 6, on circumlunar flights, whipping around the moon and coming back home. It wouldn't be a lunar landing—it wouldn't even be a lunar orbit—but it would still be a first-ever trip to the moon and back, meaning that the first eyes ever to see the lunar far side would be Russian. Yes, for Zond 5 at least, they would be mouse or bug or hamster eyes. Still, assuming that the animal passengers survived the trip, cosmonauts would follow, quite possibly before the end of the year.

That got Low's teeth gnashing and his thoughts churning, which, in turn, led to his brainstorm. The decision to share his risky plan first with Gilruth and Kraft was shrewd. The two men were at just the right level on the NASA leadership trellis: they were low enough that their work still involved creative problem solving and flight planning but high enough that they were part of the administrative elite that helped make the final decisions. That meant their opinion would carry as much weight as Low's own if they decided to pitch the idea to Jim Webb, the NASA administrator.

Frank Borman in 1951, during the dreariest assignment of his career: chief of road maintenance at an American military base in the Philippines. Borman had been set to fly fighter jets in Korea, but a ruptured eardrum suffered during a training flight temporarily grounded him.
[courtesy of Frank Borman]

In early 1965, command pilot Frank Borman and pilot Jim Lovell were chosen to fly Gemini 7. On December 4, Borman (right) and Lovell emerged from the suit-up trailer and headed to the launchpad; they would spend what was then a record-setting two weeks in orbit.

Command pilot and space veteran Wally Schirra and rookie pilot Tom Stafford were assigned to Gemini 6. In October 1965, Schirra (seated) and Stafford went through suit checks in preparation for their mission later that month. At the last minute their flight would be delayed.

Gemini 7 was photographed by Gemini 6 from a distance of just thirty-seven feet on December 15, 1965. The two ships eventually drew to within a single foot of each other, flying in perfect formation at matching speeds of 17,500 miles per hour.

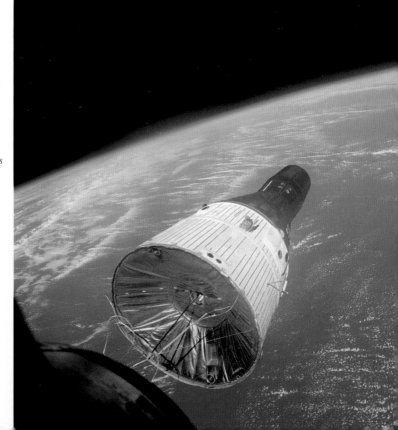

Chris Kraft, NASA's director of flight operations, oversaw nearly every step of every mission during both planning and execution. Here he is seen at his administrator's console at the back of Mission Control during the flight of Gemini 4 in 1965.

Gene Kranz was one of the rotating team of flight directors who supervised all the other controllers in Mission Control; during his shifts, he called the moment-to-moment shots. Here he is seen during a simulation session before the Gemini 4 mission.

The first Apollo mission was scheduled to take place in February 1967. Nearly a year before the planned launch, the ship's crew—space veterans Gus Grissom (left) and Ed White (center), along with rookie Roger Chaffee—posed in front of the launchpad. They would never make the trip.

Borman, a veteran of just one flight, was named to lead NASA's team overseeing the redesign of the Apollo spacecraft. He spent much of 1967 and 1968 on the factory floor in Downey, California, where the Apollos were being built.

The giant Saturn V rocket that would carry Apollo 8's crew into space stood 363 feet tall and produced 7.5 million pounds of thrust. Here, on the night before liftoff, the rocket stands illuminated on its Cape Canaveral launchpad.

The Apollo spacecraft was actually a two-part ship: the conical command module, which housed the crew, and the cylindrical service module, where the ship's fuel tanks and life-support systems were located. The spacecraft's giant engine bell protruded from the

George Low (left) was the manager of the Apollo Program Office during the run-up to Apollo 8. Wernher von Braun, who had served the Nazi regime during World War II, was the chief designer of the Saturn V rocket. Here the two men confer in the Cape Canaveral launch room.

Robert Gilruth was the director of the Manned Spaceflight Center in Houston, home to Mission Control, during the Apollo 8 mission. In 1962, he presented a model of the imagined Apollo spacecraft to President Kennedy.

James Webb, administrator of NASA, would make the final decision to approve the Apollo 8 mission, and it wasn't an easy call. Here, he testifies before Congress about the Apollo 1 fire, a tragedy that haunted every choice he would later make. [Bettmann/Getty]

Admiral John McCain Jr. was commander of the Navy's Pacific Fleet from 1968 to 1972. McCain would decide whether a sufficient recovery fleet could be called back from a brief Christmas vacation to retrieve the Apollo 8 spacecraft after splashdown. [Bettmann/Getty]

Before Apollo 8 could be cleared to fly, Apollo 7—the earth-orbital shakedown mission—would have to succeed. The crew was (left to right) rookie Donn Eisele, veteran Wally Schirra, and Walt Cunningham, another first-timer. Deke Slayton (far right) spent months preparing the crew for the mission. [Getty]

The Apollo 8 crew (left to right): Jim Lovell, the command module pilot; Bill Anders, a rookie pilot; and Frank Borman, the mission's commander. This photograph, a NASA publicity shot, was taken on the steps of the Apollo simulator five weeks before launch.

The Borman family, photographed in the months before Apollo 8's launch: Frank (right), his wife, Susan (seated), and their sons, Fred (left) and Ed
[ullstein bild via Getty Images]

The Lovell family: Jim (far right), his wife, Marilyn (far left), and their children (clockwise from top), Jay, Barbara, Jeffrey, and Susan.

The Anders family: Bill (top left), his wife, Valerie (center), and their children (clockwise from top), Gregory, Alan, Glen, Gayle, and Eric.

On Apollo 8's launch day—December 21, 1968—Borman suits up before the ride to the launchpad. Here a technician is tucking a flight pen into an arm sleeve while Lovell sits in the background.

Lovell also got help with his pen as he was suited up. The Apollo 8 mission patch—a figure eight circling the Earth and the moon—was Lovell's idea. It roughly tracks the trajectory the mission followed, though it doesn't show that the spacecraft orbited the moon.

Anders suiting up on launch day. Underneath their bubble helmets, Apollo astronauts wore communications headgear that became affectionately nicknamed Snoopy hats, after the cartoon dog.

CBS's Walter Cronkite covered Apollo 8's launch from a makeshift newsroom in Cape Canaveral and then returned to his newsroom in New York for the rest of the mission. Here he stands before a diorama at New York's Hayden Planetarium two weeks before the launch. [CBS Photo Archive/Getty Images]

The Cape Canaveral firing room on the morning of the Apollo 8 launch, with the video screens showing the time until liftoff. The moment the rocket cleared the launch tower, Mission Control in Houston would take over.

An estimated 250,000 people jammed the shorelines and campsites around Cape Canaveral to witness Apollo 8's launch. The license plates indicated that they had driven in from as far away as Canada. This photo was taken shortly after liftoff.

Borman, Lovell, and Anders (foreground to background) fully suited and preparing to head for their rocket on launch day. From the moment their helmets went on, the astronauts were breathing suit air, and they would not taste earthly air again until they splashed down six days later.

At 4:42 a.m. eastern time, the Apollo 8 crew walked from the suit-up building to the launchpad van. Liftoff was set for 7:51 a.m. Despite the early hour, swarms of journalists, kept behind barricades, showed up. Inside the astronauts' bubble helmets, it was all a muffled roar.

Lovell, assigned to the center spot in the cockpit, was the last of the three astronauts to climb into the spacecraft, since the hatch was directly above his seat. The spacecraft was surrounded by a so-called White Room at the end of a swing arm at the top of the gantry tower.

Apollo 8 lifted off precisely on time, the titanic force of its engines producing vibrations that were detected on seismographs as far north as New York.

Apollo 8's first-stage engines burned for two minutes and forty-one seconds. By the time they shut down and the stage was jettisoned, the crew was nearly halfway to Earth orbit and moving at more than 6,100 miles per hour.

Just over half an hour after blasting out of orbit, Apollo 8 had already traveled twelve thousand miles. At that point in the flight the crew jettisoned the spent third-stage engine—the last remaining part of the Saturn V—and then turned around and photographed it.

The farther Apollo 8 traveled from Earth, the more the crew could see of the home planet. A shot taken during the early portion of the voyage—in which the planet appears to be upside down—shows West Africa (bottom left) and a vast stretch of the Atlantic Ocean

Borman, captured in a frame from the 16-millimeter onboard movie taken by the Apollo 8 crew

Lovell, sighting stars at his navigation station, in a frame from the 16-millimeter movie

Anders, in a frame from the same footage, showing a generous growth of beard, which suggests that the image was captured more than halfway through the mission

(opposite left) Anders—the mission photographer—took this shot of the far side of the moon during one of Apollo 8's lunar orbits. The spacecraft was then soaring over the Southern Sea; for scale, the dark crater to the right of center is about 43 miles in diameter.

(opposite right) Anders photographed the Sea of Tranquillity in high relief, and the angle of the sun made the moonscape look more jagged than it actually was. Tranquillity's relative smoothness led NASA to select it as the site of the first lunar landing, which would occur eight months later.

(opposite) During the third of Apollo 8's ten orbits, the angle of the spacecraft's prow allowed the crew to get their first glimpse of the Earth rising over the lunar horizon. Anders captured the iconic image that would come to be known as "Earthrise."

After Apollo 8 landed in the Pacific Ocean in the predawn hours of December 27, 1968, the crew was evacuated by helicopter. The USS Yorktown then steamed to the site of the splashdown and winched the spacecraft onto the deck.

Borman, Lovell, and Anders emerge from the rescue helicopter on the deck of the Yorktown. During the short flight to the ship, Borman shaved with an electric razor; after taking a ribbing for the patchy scruff he'd grown during the Gemini 7 flight, he wanted to be clean-shaven.

Below deck aboard the Yorktown, the Apollo 8 astronauts were greeted with a welcome-home cake. The massive confection was about the right size for a crew of seventeen hundred sailors plus their three celebrated visitors.

Apollo 8's astronauts were feted in Washington on January 9, 1969. President Lyndon Johnson, who had just eleven days remaining in office, received them.
[AP Photo]

Lovell, flanked by Borman and Anders, waves during the crew's address to a joint meeting of Congress on January 9. Behind them were Vice President Hubert Humphrey (left) and Speaker of the House John McCormack.
[AP Photo]

After leaving Washington, the crew flew to New York City, where they were welcomed the next day with a tickertape parade on lower Broadway. An estimated two hundred tons of confetti fluttered down on them as spectators jammed the parade route despite the 28-degree cold.
[AP Photo]

*Before the Apollo 8 mission,
human eyes had never seen the
Earth as a whole, floating sphere.
In this photograph, the crew
captured an image of most of
the planet's western hemisphere;
West Africa is visible at the
bottom, and an inverted South
America is at the top.*

Once Low satisfied himself that his notion wasn't completely unhinged, he called Kraft and summoned him to his office. Then, revealing nothing, he asked Kraft to accompany him to Gilruth's office.

When they got there, Low made his case. He wanted to go to the moon on a circumlunar flight. He wanted Apollo 8 to make the trip. He added the condition that the lunar mission would be approved only if the Apollo 7 Earth orbital mission was "very good, if not perfect." He closed with a final requirement: If they went ahead with the moon trip, it had to happen before the end of the year, or in roughly four months. There was simply no more time to waste.

Low was met with silence, which was pretty much what he had expected. Kraft, known by both of the other men for being all but unshockable, seemed shocked at last.

Hoping not to spook his colleagues too much, Low had deliberately used the term "circumlunar," as opposed to the less specific "moon flight." A circumlunar flight, like the one the Russians were apparently planning, would keep the crew on what was known as a free-return trajectory, meaning that even if their service propulsion engine—SPS in the engineers' shorthand—failed them completely on the way out, lunar gravity would whip them around the far side of the moon and toss them home. Actually orbiting the moon, however, would require the SPS to work perfectly, not once but twice: the first time to settle the ship into lunar orbit and the second to blast it out.

Gilruth was reasonably certain that Low couldn't be suggesting that brand of insanity. But just to be certain, he checked.

"What kind of mission?" he finally asked. "Just out and back? A circumlunar trip?"

"Yes, that's what I'm thinking," Low assured him quickly.

When Kraft at last found his voice, he couldn't say much that was terribly positive, even about such a straightforward trip to the moon. "I have to think about it awhile," he said. "I can't give you an answer. We had all the stuff laid out according to our game plan, and that ruins our game plan." He shook his head. "We'll have to go back and look at everything."

That was exactly what Low had expected Kraft to say, and he was

reasonably confident that once Kraft did look at everything, he would realize that at least some of the pieces for the revised mission were already in place. The command and service module was more or less ready, which was no small accomplishment. Like the LEM, it was a two-piece vehicle. The command module part was an eleven-foot-tall conical structure that housed the cockpit in which the crew lived. The service module was a cylindrical, twenty-four-foot structure that was attached to the rear of the command module like the trailer behind the cab of a truck. It contained the power and life-support systems, as well as the machinery for the SPS. The engine's bell extended another twelve feet behind the service module.

The hard work that had been done since the fire had made it possible for NASA to certify the command and service module, or CSM, for flight. But that was only the hardware. What had not yet been finalized was what the computer guys called the "software"—the programming for the electronic brains both in the spacecraft and on the ground.

The Saturn V, meanwhile, had lost the confidence of a lot of people and would somehow have to be proven flightworthy. And the mission controllers—the people who would actually be managing the flight—had not yet begun intensive rehearsals for a moon journey they all assumed was at least a year away.

Then there was the matter of training a crew, one that could be made ready to fly in so short a time. That not inconsiderable detail was on Gilruth's mind, too. "Deke needs to be a part of this conversation," he said, picking up his phone and dialing Slayton's extension.

The chief astronaut arrived and sat. Low presented the plan he had just described to Gilruth and Kraft; betraying nothing, Slayton chewed on it for a moment. As an astronaut—albeit one who had never flown—he preferred ambitious risk taking to methodical plan making, but as the man charged with the safety of the astronauts who actually had flown and would fly some more, he had to be more cautious.

"I think I could train up a crew," he said. "That's not a concern. But as to the idea itself . . ." He trailed off. "I need to think about this carefully for a couple of days."

Low nodded, and then he and Gilruth sent Kraft and Slayton off to

begin working out the details. Just as they were standing to leave, though, Gilruth gestured for them to wait.

"Tell no one about this," he said, "no one except the people who absolutely must know. And whatever you do, say nothing to the press."

The two men agreed, and when they reached the hall, Kraft looked at Slayton. "You go do your thing," he said. "I'll go do mine."

☆ ☆ ☆

For Slayton, there wasn't much to do; nor was there anyone to consult—except himself. Whatever decision he reached would rest almost entirely on his exhaustive knowledge of the training and the aptitudes of his Apollo crews. He would not reveal anything to the astronauts themselves, all of whom would surely campaign for the circumlunar plan. He alone would decide if there was a team of three ready to make the flight.

At the moment, the two crews best prepared to fly were the two whose missions were closest to launching: Apollo 8's Jim McDivitt, Dave Scott, and Rusty Schweickart and Apollo 9's Frank Borman, Jim Lovell, and Bill Anders. Of the two, McDivitt's crew was readier, but in some respects that fact worked against them. Unlike Borman's crew, whose training had thus far been largely limited to the command module, McDivitt's men had already begun rehearsing in their LEM, even though the vehicle wasn't flight-ready yet. Switching that crew to a mission that didn't involve a LEM would mean a lot of wasted training.

The makeup of the two crews also warranted consideration. Slayton held McDivitt and Borman in equally high regard as pilots, yet there was no denying the powerful impression Borman had made on so many people at NASA with his handling of the Apollo 1 investigation and its aftermath. What was more, while none of the three astronauts on McDivitt's crew had flown together before, Borman and Lovell had spent fourteen days together on a mission no one else had wanted—and they'd flown it well and in the process had forged a unique piloting partnership. If it would be a shame to waste McDivitt's crew's LEM training, it would be at least as much of a shame to waste the crew cohesion Borman and Lovell could bring to the first lunar flight.

Slayton would spend a few days thinking it all through; his committee of one needed a bit more reflection before he announced his decision. But if he had to make the call today, he knew what it would be.

For Kraft, there could be no such solitary contemplation. He needed a real committee—a small, trusted group of deputies—who could work with him behind tightly closed doors.

The moment he returned to his office, he summoned a handful of colleagues, including two key lieutenants: Bill Tindall, an orbital mechanics expert who would effectively manage the physics of any lunar trip, and John Hodge, the director of flight planning, who would write the mission script. The team arrived, and Kraft told them what Low had proposed. He was reassured to see that none of them responded with wide eyes or a flat no. Indeed, all of them seemed pleased, even excited, by the idea.

Kraft's group began talking it through, especially the question of the timeline, which would require them not just to do something they had never done before but to do it in a sprint. Yet the more they talked, the more they realized that although the obstacles were considerable, none of them seemed insurmountable. Kraft, satisfied with what he was hearing, sent the team out to discuss it further—*among themselves only*, he reminded them—and told them to return the following day with an answer that was a product of more detailed reflection and a decent night's sleep.

When Kraft's deputies trooped back into his office the next day, the answer they provided was not at all what he had anticipated.

"We think it's a great idea," Tindall said. "We don't know whether we can do it or not, but it's worth a try." Kraft nodded, but Tindall wasn't through. "There's one thing we want to do which you haven't mentioned and you're probably going to be upset about. We don't just want to go to the moon. We want to go into orbit around the moon."

Now Kraft was truly at a loss. Looking at the other engineers, he fulfilled his most important responsibility, which was to ask the critical question: "Why would you want to increase the risk?"

His men were ready with an answer. For starters, flying to the moon was a huge commitment of resources and training, and flying there just to whip around once, take a peek at the lunar backside, and then return home seemed like a terrible waste. Yes, going into lunar orbit would mean that the SPS would have to work perfectly, but the engine would have to do that on any trip to the moon, and it would get a thorough shakedown on Apollo 7 first. If you couldn't trust the machine to do its job, they argued, you never should have built it in the first place.

Kraft and the administrators were uneasy about the reliability of the SPS, partly because they had been thinking too hard about one of the statistics NASA had lately taken to tossing out to the media. There were 5.6 million separate parts in the command and service module, the agency would say, which meant that even if everything functioned 99.9 percent perfectly, 5,600 parts might go bad. That shocking number was intended mostly to wow the public with the exacting nature of the work the space agency was doing, but NASA's engineers knew that the statistic was meaningless, since you would never send men into space in a machine with a performance rating of 99.9 percent. Three or four nines on the right-hand side of that decimal point was the minimum threshold for good design. Just as important, one of the reasons that the CSM had millions of parts was that many of those parts were redundant, backups or even backups to backups, and they would never be used unless the part they were supporting failed.

NASA's engineers felt confident in the big engine for another reason. The fuel used to power it would not be the usual liquid oxygen mixed with an explosive like kerosene or liquid hydrogen, which required an ignition system to get everything burning. Instead, the SPS would use what were known as hypergolic fuels—in this case hydrazine and nitrogen tetroxide—which needed merely to be brought into each other's presence inside a combustion chamber and they would burst into explosive, propulsive life. And to make sure that the fuels had no trouble moving where they needed to go, there would be no breakdown-prone pumps anywhere in the main engine's systems. In the SPS, the hydrazine and tetroxide would be pushed along by pressurized helium fed into the lines.

If the engine was reliable enough for a lunar-orbit mission—and Kraft's engineers believed it was—there was another excellent reason not to limit the mission to a circumlunar one. America's unmanned lunar spacecraft—its crash landers and soft landers and orbiters—had long been flummoxed by lunar gravity. The moon's gravitational field had bumps in it, unpredictable areas of more than the usual one-sixth of a g, and they could potentially send a ship off course. The irregularities were the result of mass concentrations—or mascons—the remains of heavy-metal meteorites that had long ago crashed into the moon and buried themselves there. The mascons had been entirely harmless for billions of years, but as soon as a machine from Earth appeared overhead, they would reach up with their invisible gravitational hands and shake the spacecraft in who knew what ways.

"Every time we plan trajectories with the gravitational models of the moon we have, we miss by two miles," Tindall told Kraft. "But if we could put the command module in orbit, we could develop an empirical set of formulas that would help us do the right orbit prediction." Ten orbits seemed like a sensible number of circuits to make around the moon. That would allow for the collection of a good mascon data set, not to mention some thorough photographic reconnaissance of the surface that could be used to plan later landings.

This argument, Tindall knew, would carry a lot of weight with his boss. Though a lunar-orbit mission meant increasing the risk for Apollo 8, it also meant decreasing it for every lunar Apollo that would follow. It was simple arithmetic, and Kraft loved arithmetic.

After talking it through for a bit longer, Kraft dismissed Tindall and the rest of his team to go run more numbers and do the ciphering that would prove they could accomplish their part of the planning. That would take a couple more days, which Kraft believed he and everyone else needed.

Finally, at the end of the week, he summoned the one other person whose opinion he most needed: Gene Kranz. Kranz had become such a critical figure in the overall operation of Mission Control that he no longer had the time to man a console on every mission. Instead, he was working every other flight, serving as flight director only on odd-numbered missions; as a consequence, he wouldn't be chained to a control room chair for Apollo 8. This was the best possible news for Kraft, since he needed Kranz to take a more sweeping hand in planning for the mission and determining whether it could be flown at all.

When Kranz arrived, Kraft got straight to business.

"What we're going to do is propose to go to the moon this December," he said simply. Kranz said nothing, and Kraft went on: "I want you to go back and tell me tomorrow morning who should work up this plan and if you see any reason we shouldn't do this."

But Kranz had no interest in waiting until the next morning, so he borrowed Kraft's phone and called Bob Ernal, his chief computer man. Ernal hurried to Kraft's office and learned what his bosses were cooking up. When Kraft asked Ernal if his computers could handle such a job, Ernal was succinct in his response.

"Chris, if you'll give me every computer in Building 12 and Building 30 for the entire weekend, I will give you an answer on Monday."

Ernal got his computers and got his weekend and on Monday reported to his higher-ups. Yes, he said, it could be done. If all of the other parts of the operation were in place for a launch, his computers would put Apollo 8 into orbit around the moon—and, more important, they would get the ship home again.

☆ ☆ ☆

Bit by bit, system by system, the levers of the great American moon machine were being thrown, the gears meshing and beginning to turn. But there was one final part to address, the biggest, loudest, most powerful one of all: the Saturn V. If there was no rocket, there could be no moon trip.

No sooner had the good word came back from Ernal than Gilruth met with Low and Kraft once more and told them, "Before we commit to this, we're going to have to get Wernher on board." He picked up his phone and called Wernher von Braun's office at the Marshall Space Flight Center in Huntsville, Alabama, where the Saturn V was designed and parts of it were built.

The chief designer's secretary answered. "He's in a meeting," she said. "Can he call you back?"

"No," Gilruth said. "He needs to come to the phone now."

The surprised secretary went off to collect von Braun. When he picked up the receiver and said hello, Gilruth told him that he, Low, and Kraft needed to come talk to him.

"I can probably see you tomorrow," von Braun demurred.

"No, I want to see you right now."

"What do you mean?"

"We'll fly over there right now and talk to you," Gilruth answered, concluding the call before von Braun could object.

Kraft looked at Gilruth and could not completely contain a smile. He was, he knew, in a cowboy business, and he tried never to let the thrill of the work color his decisions. But today he simply couldn't help it. He phoned the commissary for sandwiches, grabbed them to go, and then he, Gilruth, and Low hopped aboard the Gulfstream plane they used to fly from Houston to Huntsville or Canaveral. Kraft hated the Gulfstream. The space agency might be launching rockets to the moon, but its budget didn't allow for a jet for the administrators. That meant tolerating the head-poundingly loud Gulfstream propellers every time they needed to fly somewhere. Even now, at just forty-four years old, Kraft could tell that the plane was wrecking his hearing. Today, however, he would bear it.

When the three men got to Huntsville, they went straight to von Braun's office but did not sit. Instead, they got him up, steered him to a conference room, and closed the door—the better to talk in complete privacy. They shared their new plan for Apollo 8 and asked him if his rocket could be ready in time. Von Braun's face told them his answer. He had a near-paternal love for the Saturn V; just as a father's hopes can become entangled with the achievements of his son, so von Braun's ambitions rode on his machine. The idea of going to the moon before the end of the year was irresistible. His first Saturn V had flown brilliantly; his second had mostly botched the job. His third would make no such mistakes.

"Yes," the chief designer told his visitors. "Yes, we can be ready in time."

Then came the final hurdle—three hurdles, actually, and none of them had anything to do with machinery. All the moon talk would amount to nothing if the plans for Apollo 8 didn't get the agreement of Jim Webb, the NASA administrator; George Mueller, his deputy for manned space flight; and President Lyndon Johnson. Webb and Mueller were attending an international conference in Vienna, and when Low phoned them overseas and at last told them that their engineers in Houston had been discussing the idea of revising the mission, they were furious. Low had never intended to bring his two bosses in on the plan until he was certain that a lunar mission was feasible, the better to reassure them that the idea had been rigorously thought through. But making the call when they were out of the country had just the opposite effect, creating the impression that the moment the bosses' backs were turned, the agency had run riot.

"Can't do that!" Mueller snapped. "That's craziness!"

Webb was even angrier. "You try to change the entire direction of the program while I'm out of the country?" he asked, incredulous. Webb was not a man to discourage ambitious thinking, much less call it mutiny, but what Low and his gang were up to came awfully close. Still, he understood the reasoning behind the plan and agreed not to rule it out until he returned home in a few days. That would be on August 22, and when he arrived, he wanted to meet with the entire team of lunar plotters.

The day arrived, and Webb and Mueller came to Houston willing—but not remotely expecting—to be persuaded by the pitch they were about to hear. In the meeting that followed, the same scene that had unfolded over and over in Houston during the preceding weeks was repeated: one or two deeply skeptical men listened quietly and carefully and then slowly became swept up in the ambition and vision and sheer brass of the plan they were hearing. And just like the people in those earlier meetings, they came away agreeing that it was time—and there was reason—to take a very big risk.

Mueller proved to be a harder sell than Webb, but Webb's vote was the one that really counted. The administrator reasoned that he was bound by a promise he had made to himself in November 1963, after the president who had hired him to run the nation's space program had been slain: he had pledged that he would get America to the moon before 1970, even if it meant throwing the dice now and then.

Yes, Webb agreed, he would take this plan to Johnson, along with his recommendation that the president approve it. He had no doubt that Johnson, with his deep feeling for the space program, would quickly agree. And he was right.

☆ ☆ ☆

The phone call that interrupted Frank Borman, Jim Lovell, and Bill Anders at the North American Aviation plant in Downey, California, was an unwelcome distraction. They had less than seven months to get ready for their Apollo 9 launch, they hadn't had nearly enough practice time inside their command module, and Anders had barely gotten his hands on his LEM at all.

Still, when the technician who came to fetch Borman told him that Slayton was on the line, the call could not be refused. Borman muttered a "Carry on" to Lovell and Anders and hurried off. A few minutes later, he came back only to say that he'd been ordered to Houston but had no idea why. Not until the next day, when he returned to Downey, did he tell his crewmates what Slayton had told him.

"We're flying early, in December," he reported, dropping the smaller news first. Lovell and Anders looked mildly surprised. "And we're going to orbit the moon."

Surprise turned to something approaching shock, but both astronauts were clearly delighted.

"Whose idea was this?" Lovell asked incredulously.

"Deke's, Kraft's—everybody's really," Borman answered. "They want it to happen, and they want it before the end of the year. We have sixteen weeks to train."

"Tight, but we can do it," Lovell said, and Borman nodded in agreement.

"But the LEM's not ready," Anders pointed out.

This was the part Borman had dreaded. "No LEM," he said. "Just the CSM."

Anders's face fell. Then he rearranged his features and forced a smile and a nod. Nevertheless, his disappointment was unmistakable. The moon was fine—wonderful, actually—but losing his LEM was like losing a limb. A man could be on a command module track or a lunar module track. He had been the front-runner on the LEM track, but now someone else would take his place, someone who would spend the next sixteen weeks continuing to train to fly the lander while Anders switched his attention to the less complex, less fanciful command module.

Still, there was no denying that serving as a crew member on the first trip to the moon would be truly historic. And part of Anders's new assignment, as Borman now explained it to him, would be to manage photographic and eyeball surveillance of possible lunar landing sites, a job that would take advantage of Anders's specialty in geology. This would be a critical part of the mission, since nobody could yet say with certainty whether the smooth plains—the so-called seas—on the surface of the moon that were being targeted for future missions were the pristine landing sites they appeared to be or whether boulder fields or other irregularities would menace a LEM as it made its final approach. What's more, flying so early in the lunar-mission cycle might well give him a chance at another flight, this time in the commander's seat. On balance, he guessed that what Borman had told him today was good, if not ideal, news.

Anders would still have to break the surprising news to his family. Borman mentioned that he had told Susan and the boys about the new plan already, when he was back in Houston, and that they had given their approval without hesitation or even much excitement. Anders expected he would get much the same reaction out of his brood.

He had married his wife, the former Valerie Hoard, in 1955, and they now had five children—four boys and one girl, ages four through eleven. Valerie had known Bill only as a Navy man, first setting eyes on him when she was just sixteen and he was a nineteen-year-old naval academy plebe. They had met improbably, when he had traveled from Annapolis to his boyhood home of San Diego for a holiday and asked a local girl out on a date. The girl had accepted, but only on the condition that a friend of hers named Valerie could come along. Anders agreed, and soon he was very glad he had, since he took to Valerie more than he did his intended date—and Valerie took to him. As was so often the way with service academy couples in the full flush of love, Bill and Valerie married just after his graduation.

An aviator's wife before she was even old enough to vote, Valerie had long since grown accustomed to accepting her husband's flying assignments with equanimity, and a mission to the moon would be no exception. The children were equally unmoved. Growing up on an astronauts' block and going to a school with other astronauts' children, even the youngest ones knew little else but the life of an astronaut's child.

"You know what really impresses the kids?" Anders would ask friends when they seemed a little too awed by his job. "When one of the other kids' fathers is a fireman."

Only Lovell got much of a reaction from his family when he told them he was going to the moon. He and Marilyn had been planning to take their two boys and two girls—who were now fifteen, thirteen, ten, and two—to Acapulco for the Christmas holidays, and Marilyn had already begun shopping for clothes and beach supplies for the trip. But NASA had now filled in the Lovells' holiday calendar for them.

"You know that Acapulco trip we're planning?" Lovell asked Marilyn when he came home from California.

"I do," she said warily.

"Well, I'm thinking of going somewhere else instead."

"And where would that be?" Marilyn asked.

"The moon," he answered.

Marilyn, despite herself, smiled.

EIGHT

ONCE NASA DECIDED TO GO FORWARD WITH A JOURNEY TO THE moon before the year was even out, the Apollo 8 astronauts, mission controllers, and flight planners had to move fast. They had only sixteen weeks before launch day, and now the clock hands counting off that time began moving very quickly. Wernher von Braun, as much as anyone, was quite open about how urgent the job was, and he found an evocative way to express it, even though English was not his first language.

"The moon in the sky," he would say, "has become a deadline display device." And unlike the countdown clocks at Cape Kennedy, it was a device that nobody could ever turn off.

For von Braun, the sixteen weeks would have to be front-loaded, because although the Saturn V would play the most short-lived role in the lunar enterprise—less than three hours after liftoff, the rocket's 363-foot stack would be reduced to just 35 feet of command and service module— it was decidedly the most difficult component to prepare and ship and configure for launch. And this time, the Saturn had to fly true. If it went pogo again, it might shake the astronauts to death. If it failed catastrophically, it would unleash a monstrous and inevitably lethal explosion.

Immediately following the disastrous flight of Apollo 6, von Braun had gathered his entire Huntsville team to fix the problems that had almost wrecked the ship. Just to make sure he had all of the smartest heads together in the same room, he had also summoned—either in person or by phone—nearly one thousand other engineers from every contractor, subcontractor, and sub-subcontractor who might have a useful thought about how best to sort out the giant machine's multiple defects.

The pogo problem, the most critical issue, was traced to a simple fact that is true of all rockets: they burn fuel, and as that fuel vanishes, it leaves nothing behind. The lower the level of kerosene, liquid hydrogen, and liquid oxygen in the tanks, the less mass there was to absorb the rocket's vibrations, which, in turn, had led to the pogo oscillations.

The answer was to keep that emptying space full—but with what? The Saturn V's engines had only one speed, which was full speed, and the only way for the rocket to accelerate continually as it climbed was for it to get lighter and lighter as its fuel burned away, resulting in a fixed thrust that was propelling an ever-dwindling mass. That's how you got from zero miles per hour to the nearly 25,000 miles per hour it would take to get to the moon. The problem was, if you added something to the tanks to fill the empty spaces, you would be adding weight, too.

The engineers solved the puzzle neatly with helium. It was inert and it was light; in fact, at atmospheric pressure, the stuff actually had *negative* weight. That would not be the case in the Saturn V, since the helium would be pressurized and thus more densely packed, but it still weighed effectively nothing. All you had to do was steadily pump helium into the tops of the tanks to take the place of the fuel as it burned away, and the gas would absorb the launch oscillations like a properly inflated tire cushioning the bumps in a road.

The failure of the two engines on the second stage was a knottier issue. Pogoing was part of the problem here, too: during the flight of Apollo 6, it was so severe that it had caused the I-beam deflection. But another part was simple sloppiness—potentially deadly sloppiness—in the factory, where the electrical leads to the two engines had been crossed. The wrong power systems were thus feeding the engines, meaning the systems couldn't respond properly when one engine or the other called for more or less power. That boneheaded error was quickly exposed in the post–Apollo 6 investigation, and von Braun had been assured that there would be no such nonsense going forward.

These and other fixes and refinements gave von Braun confidence that the next Saturn V would be fit to carry a crew. And this made him confident that the Apollo 8 crew could go to the moon.

"Once you put a man on it, it doesn't matter where you send it," he said in the days after Gilruth, Low, and Kraft came to visit him in Huntsville.

NASA itself wasn't yet convinced, however. Before the agency would approve von Braun's rocket for a manned flight, the chief designer would have to make his case to Dieter Grau.

Like von Braun, Grau had been one of the German rocket engineers who'd come to the States after the war. He had worked for von Braun in Germany, and now he worked for him in Alabama as well. But Grau was also Huntsville's chief of quality and reliability operations, and in that capacity he would, at certain moments, be his boss's boss. This was one of them: von Braun could argue all he wanted that his rocket was ready for flight, but it wasn't going anywhere until Grau gave it the okay.

"What more should be done?" von Braun asked him in September, after he had done all the work he could do on the rocket.

"I want the opportunity to do one more complete check," Grau answered.

Von Braun agreed, and over the course of the next few weeks, the quality and reliability team went over the schematics and design history of every element of the Saturn V that would carry Apollo 8, as well as the performance telemetry from Apollos 4 and 6 and the contractors' records of the testing they had done after those earlier flights. Precisely as Grau had feared, several little problems presented themselves—none critical, perhaps, but none acceptable. Only when every component in every system in the rocket had been set right and certified fit did Grau give the machine his approval.

Even then it was a conditional approval. Once the rocket was on the launchpad at Cape Kennedy, it would be out of Huntsville's hands, but there was plenty of proxy hardware still in Alabama. Grau ordered up continued firing tests on Saturn V engines straight through the fall.

The final tests would take place in December, with the last one scheduled for the eighteenth of that month, just three days before Apollo 8's launch.

☆ ☆ ☆

Creating the necessary software and writing the needed computer code was another hurdle for the Apollo 8 team. Gene Kranz might have given Bob Ernal the use of every computer in Buildings 12 and 30 for an entire weekend, but a thumbs-up from Ernal after just two days of work was hardly the final answer. It was, in fact, only the signal that it was time for a lot more people to get together and do a lot more work. For this, Bill Tindall would again play a role.

Tindall, who had played a central part in the debate by Chris Kraft's team about whether Apollo 8 should orbit the moon, was known by Gene Kranz and other higher-ups at NASA as "the architect"—a hat tip to his status as the person who had blueprinted all of the Apollo's computer programs. To his subordinates, he was known for his blunt, sometimes blistering technical memos, which were unsparing when the architect wanted to make it very clear just what he thought of this or that piece of software built by one of his team members.

As August turned to September, with the sixteen-week clock ticking, Tindall took to spending more and more of his time in Cambridge, Massachusetts, at MIT's Draper Lab, where the mission's software was being written and tested. Before long, four hundred engineers were working at MIT under the nominal authority of the university but the practical authority of Tindall. The spacecraft's memory was divided into thirty-eight separate banks, each capable of storing a full kilobyte of information—or one thousand discrete units of data. That was a lot of computing power to pack into a small space.

The focus of most of their work was the Colossus software, the system that—after winning a years-long competition with other programs—had been chosen as the one that would be used to navigate to and from the moon. But Colossus wasn't completely ready, and while the pace of the work to make it fit for flight had already been fierce, now it would have to move at a full sprint.

The most serious concern was what to do in the event of software crashes, which occurred often and might never be entirely eliminated from the program. The key to responding to crashes was how quickly and seamlessly the system could reboot. On a three-day outward coast to the moon, you might be able to tolerate it if your computer went off-line for a little while, but in the critical minutes or seconds before the engine burn that would cause the spacecraft to enter lunar orbit, a software crash would be disastrous.

The work at MIT was thus focused on ensuring that the restarts following a crash would be fast and automatic. The computer's core operating program would have to know which systems to start up in which sequence, how to check the soundness of each system as it came online, and how to suspend all other nonessential operations aboard the spacecraft while that work was being done.

Tindall drove the team hard and was not above using the engineers' own, often grand, egos against them. If MIT couldn't handle the job, he would casually muse, he could always call in some hotshots from IBM for consultations. The newly motivated MIT engineers would promptly assure him that no such help would be necessary.

As the fall wore on, the bugs got written out of the software. When Colossus was finally considered ready for flight, nobody could say with certainty if that confidence was well placed. But Tindall pronounced himself happy, and that counted for a lot.

✳ ✳ ✳

Assuming the Apollo 8 spacecraft did get off the ground and did make it to the moon and did get back to Earth safely, there was still the business of rescuing the astronauts after splashdown. Technically, it wasn't a difficult problem: you sent out ships and plucked the crew out of the water, and in the seven years Americans had been flying in space, NASA and the Navy had gotten very good at that job. Nevertheless, it was always a massive logistical headache.

There were a few possible launch windows at the end of 1968, and all of them were dictated by the relative positions of the Earth and the moon. The moon was a moving target, orbiting the Earth at a speed of 2,288 miles per hour. This meant that when Apollo 8 left the ground, it would be aimed not at the spot the moon was occupying in the sky at that moment but toward where it would be three days hence—much the way hunters direct their buckshot ahead of a flock of flying ducks, rather than at the ducks themselves.

For Apollo 8 to succeed, three conditions would have to be met. First, Earth would have to be at the precise spot in its rotation where the Saturn V could leave the ground, enter Earth orbit, and then blast out to the moon at the proper angle of approach. Second, either the Atlantic or the Pacific Ocean would have to rotate into position so that it would be underneath the spacecraft when, six days after launch, it arced through the atmosphere for splashdown. Third, the moon would have to be in the right spot in its waxing and waning phases so that the portions of the surface the crew planned to survey as future landing sites would be properly illuminated. After much calculation, NASA determined that the best of the possible windows put the liftoff at 7:51 a.m. eastern time on December 21, with the spacecraft going into lunar orbit on Christmas Eve and splashing down in the Pacific Ocean southwest of Hawaii in the predawn hours of December 27.

The location and timing of Apollo 8's splashdown required that a complex set of orders be given to the recovery ships, and some of those ships might already have commitments. The Navy was planning to give its sailors a short Christmas vacation, and the decision to keep some of them working so that they could recover the spacecraft would rest with Admiral John McCain Jr., the commander of the Pacific Fleet and of all naval forces in Vietnam. McCain had a lot of skin in the Vietnam mess; his firstborn son, naval aviator John McCain III, was being held as a prisoner of war in North Vietnam, having been shot down and captured almost a year earlier. Still, the Navy and its senior officers had myriad responsibilities, and they included supporting the nation's space program—as long as that support didn't compromise naval operations.

It was Chris Kraft's job to persuade Admiral McCain that Apollo 8 met these criteria. In October, Kraft flew out to Hawaii and addressed an auditorium full of Navy brass, though most of his attention was directed at McCain, who sat smoking a large cigar, surrounded by his officers. Point by point, the space man delivered a meticulous presentation to the Navy man about the mission's goals, its stakes, and, most important, its vaulting ambition: getting American astronauts out to the moon and back before the Russians could beat them to it.

"Admiral," Kraft concluded, "I realize that the Navy has made its Christmas plans, and I'm asking you to change them. I'm here to request that the Navy support us and have ships out there before we launch and through Christmas. We need you."

McCain did not need to give the request more than a moment's reflection. "Best damn briefing I've ever had," he said. "Give this young man anything he wants."

☆ ☆ ☆

If the teams in Huntsville and Cambridge and NASA headquarters were sprinting toward December, the men in Houston were working even harder—specifically, the three who would fly the Apollo 8 spacecraft and the dozens who would rotate through the consoles at Mission Control. And if the engineers at MIT had their Tindall to run them ragged, the people in Houston had their simsups, and they were even worse. The simsups were the simulation supervisors, the men whose job it was to make the lives of a lot of other people miserable. You could actually see the simsups at work; they sat at their own bank of consoles on the right side of the Mission Control auditorium. But you couldn't quite get to them: a wall of glass separated them from the rest of the room, which was probably just as well.

For both the astronauts and the mission controllers, the bulk of the training involved running simulated missions, then running them again and again and again so that everyone knew every step in every possible flight plan deeply, exhaustively, reflexively. And then the simsups would blow everything up. They would allow a routine rehearsal in Mission Control to run for a while, and then, with no warning, they would shut down three of the imaginary first-stage Saturn V engines when the rocket was only a thousand feet off the launchpad. Or they would kill the communications systems five minutes after the crew had left Earth orbit on the way to the moon, and when controllers would try to switch to the backup system, they would take that out, too. Or they would crash the spacecraft's environmental system and then order the man at the environmental control console to get the system configured again before the astronauts died from hypothermia or lack of oxygen.

The astronauts rehearsed in a spacecraft simulator elsewhere on the Houston campus, and the simsups were just as merciless when working with them. They would get the crew all the way to the far side of the moon, overburn their main engine, and then give them precisely three minutes to sort out the problem before their orbit irreversibly decayed and they headed for a crash landing on the lunar surface. They would send the command module into a high-speed spin halfway to the moon and then kill the thruster controls, meaning the crew would have to bring their backup systems online before they could even begin to stabilize their spacecraft—and they would have to do it all before the simulated spin rate reached sixty revolutions per minute, or one per second, at which point real astronauts in a spacecraft that was actually spinning would suffer extreme vertigo and lose consciousness.

Sometimes the simulations were run only with the astronauts or only with the mission controllers. Other times the simsups would conduct so-called integrated sims, when the people on both ends of the voice and telemetry links would play the same roles they'd take days or weeks later when the mission actually flew. That was the way astronauts and mission controllers had trained for every flight that had ever flown, and that was the way they trained for Apollo 8. Except this time they were training to fly to the moon—or, more accurately, they were learning to fly to the moon, since no one had ever done it before.

☆ ☆ ☆

Throughout the run-up to the launch of Apollo 8, it had remained an absolutely inviolable condition that Apollo 7's mission should, in George Low's words, be "very good, if not perfect," or the Apollo 8 crew would not be going to the moon.

Apollo 7's commander was Wally Schirra, and the happy, jokey Wally known to all during the Gemini era was still in little evidence. Ever since the fire, he had been replaced by the Wally who prowled and scowled on North American Aviation's factory floor, determined to prevent another disaster such as the one that killed three of his fellow astronauts. Not long before the launch of Apollo 7, he and his crew—Walt

Cunningham and Donn Eisele—sat for a press conference in Houston, and from the beginning, Schirra seemed testy, impatient, and not a bit happy to be there. Things got especially awkward when someone asked him how comfortable he was with the soundness of his spacecraft.

"We've basically lived with Apollo 7 at the plant, we've lived with it at the Cape, and if somebody takes even a small component off it, we immediately become furious and say, 'Why did you remove it?'" he responded. "We expect answers immediately."

The journalists in the room exchanged sidelong glances, while the North American executives at the meeting shifted uncomfortably. This kind of talk was always awkward, although tolerable behind closed doors at the factory. But to share it with the press? It just wasn't done.

And Wally wasn't finished.

"In fact," he went on, "I'm waiting for an answer, as an example, why someone took our hatch cover off and took it to Downey and we haven't got the answer on why."

The North American men blanched. Not the hatch, anything but the hatch—the bank-vault door that had killed Grissom, White, and Chaffee. If the engineers had removed the hatch from the spacecraft at the Cape and flown it back to California for adjustments, they'd probably had a good reason. But the mere mention of that one particular part sent a shudder through the room.

"Pardon me, I just went into shock," said Eisele, a rookie who had no business speaking out of turn. But as long as he was following the lead of the commander, he apparently believed, he could say anything he wanted.

"I know," said Schirra.

"Oh boy," answered Eisele.

At 11:02 a.m. on October 11, Apollo 7 blasted off from Cape Kennedy atop a Saturn 1B rocket, the smaller version of the Saturn that would be used for this first manned mission of the Apollo series. The booster worked flawlessly, and in the early going the spacecraft also performed well. But the crew was another matter, and the three men who had been trouble on the ground became unbearable in space. Just fourteen hours after they took off, Cunningham reported to the ground that Schirra had come down with a serious head cold. In the cramped command module, where the air was recirculated, the windows couldn't be opened, and any surface touched by one man would inevitably be touched by the others, Cunningham and Eisele quickly caught the bug, too. The crew members' bad health fouled their tempers further, and it didn't help that the people who had drawn up the flight plan—itchy after nearly two years without a manned space flight—had stuffed the schedule with so many experiments and maneuvers that the astronauts barely had a chance to rest and catch their hacking, congested breath. Before long, barely a pleasant word passed between the crew in orbit and the controllers on the ground.

"I wish you would find out the idiot's name who thought up this test," Schirra snapped at the capcom after a navigational exercise did not work as planned. "I want to talk to him personally when I get down."

"While you're at it, find out who dreamed up the horizon test, too," Eisele tossed in. "That was another beauty."

When a backup evaporator failed to work properly, potentially limiting the astronauts' water supply, engineers on the ground came up with a makeshift fix—just the kind of work-around technicians were good at devising and that a crew usually appreciated. But not this crew.

"Is this something that somebody's dreamed up after all these months?" Cunningham growled. "I've been told you can't reservice a secondary evaporator."

The ground responded that that used to be correct, but now there was a way, and it would be a relatively easy four-step fix.

"Okay, hit me with it," Cunningham answered. "It looks pretty Mickey Mouse to me, but I'll stand by if I have to do it."

Finally, mere insubordination turned to rank mutiny. On October 22, the mission's final day, it came time for the crew to don their pressure suits and helmets before reentry. This protocol had always been followed and always would be, as far as NASA safety engineers were concerned, in case the fiery plunge through the atmosphere caused a breach in the ship, leading to a sudden depressurization. But Schirra was above such things and refused to wear his helmet or to insist that his crew do so, lest their congested ears suffer from the pressure.

The capcom demanded that the rule be observed, but Schirra was unmoved. Finally Deke Slayton took the microphone and spoke directly to the spacecraft. This was unprecedented: in order to avoid contradictory commands being sent to the ship, the capcom's voice was the only one the astronauts were supposed to hear. Even Slayton, however, could not persuade Schirra and his crew to put their helmets on.

"I guess you better be prepared to explain in some detail when you land why you haven't got them on," Slayton said, finally relenting. "But it's your neck, and I hope you don't break it."

Schirra didn't break his neck, and Cunningham and Eisele didn't break theirs. But there was a price of another kind to pay.

Before he'd flown, Schirra had announced that Apollo 7 would be his third and last mission. But Cunningham and Eisele—two men who had been brave enough to be part of the first crew to fly an unproven machine—had very bright lunar futures. After Apollo 7, however, that was all over. No sooner did they land than Kraft dropped the same hammer on them that he had once dropped on Scott Carpenter. Eisele would never fly for him again; Cunningham almost certainly wouldn't either, though since, of the three, he had offended the least, his sentence included the possibility of parole. But it was a parole that never came, and before long Cunningham, too, was out of the space flight business.

The Apollo spacecraft itself, however, was very much *in* business. It was all but universally agreed within NASA that the October mission had turned out to be every bit the success Low had demanded: the command and service module had worked almost perfectly for the entire eleven days the crew was aloft. Especially important were the tests of the main engine, which repeatedly lit and shut down precisely on command. If Apollo 7's engine performed so well in orbit around the Earth, there was no reason to think Apollo 8's wouldn't do the same in orbit around the moon.

The lunar mission was on. Schirra, Cunningham, and Eisele would spend the rest of their lives earthbound, but Borman, Lovell, and Anders were going to the moon.

☆ ☆ ☆

Nobody at the Central Research Institute building outside of Moscow, where the brain trust of the Soviet space program spent their days, could spare much of a thought for any space mission the Americans might or might not be planning. They had more important matters to mind at home.

In September, more or less as planned, they had launched their Zond 5 spacecraft and whipped it around the moon with a cargo of turtles and worms and insects, bringing them within twelve hundred miles of the lunar far side. The spacecraft returned to Earth, but its aim was poor and it missed the precise atmospheric corridor needed for the high-speed reentry. It didn't miss by much, however, and although the ride was rough and the spacecraft landed in the Indian Ocean rather than on the steppes of Kazakhstan, the animals survived.

Making the mostly successful mission sweeter for the Soviets, the American surveillance ship USS *McMorris* happened to be loitering nearby when the Zond was recovered. Moscow assumed that the nest of spies aboard the ship would surely report back to Washington that the Soviet Union was about to beat them in space once more, this time with the first manned mission to the moon.

But before attempting to achieve that milestone, the Soviets would need to launch at least one more unmanned Zond flight, just to increase their confidence that a cosmonaut could survive the ride the turtles and worms and insects had. The Russian space engineers were sure they had solved the reentry problems; to prove the point, this time they would not only fly their animal passengers around the moon and back but land them a precise 9.9 miles from the launchpad. A cosmonaut who touched down that close to the pad would be practically within walking distance of the launch site barracks.

Zond 6 launched on November 10. Like Zond 5, it swung around the far side of the moon and flew straight back to Earth. Then, as it entered the atmosphere, it became clear that this time it would actually do much better than its earlier Zond brother. The ship performed the reentry maneuver almost flawlessly. It plunged through the atmosphere, building up no more heat and no more g's than a human passenger could easily handle, and headed straight for a landing at the exact spot that had been selected.

The Zond's parachute deployed when it was supposed to, and its speed of descent slowed as it was supposed to as well. Then, just 3.3 miles above the ground—after a journey of some 230,000 miles—the ship did something it was absolutely not supposed to do until it had touched the Kazakh soil: it jettisoned its parachute. There was nothing to save it then, nor would there have been a way to save a cosmonaut if one had been on board. After falling to the ground like the dead, multi-ton weight it was, the Zond half-buried itself in the earth.

In the Central Research Institute building the next day, Nikolai Pilyugin, the chief designer of the Zond's guidance system, gathered his engineers for a dressing-down. "Finally," Pilyugin shouted, "all the systems activated without a problem and you managed to shoot off the parachute when it was almost on the ground! And you were dreaming that we were about to launch a human being?"

The engineers explained that the problem had been an air leak inside the spacecraft. The leak—caused by a faulty rubber gasket—confused the instruments that were supposed to sense atmospheric pressure and, in turn, feed that information to the system that controlled the parachute so that it would know the precise moment to cut the cords. The problem could be fixed, the engineers assured Pilyugin. But they could not assure him that that fix could be made and the Zond could be test-flown in time to beat the onrushing Americans. And they could not assure him that some other problem might not doom the next Zond.

Equal parts mystified and despairing, Pilyugin could only shake his head and turn his attention to Konstantin Davidovich Bushuyev, his resident expert on flight dynamics.

"Konstantin Davidovich," he asked, "if you could please tell us, after such a good flight, why did you crash the descent module?"

NINE

Lyndon Johnson didn't need to know the names of every one of the 142 people who would be joining him in the State Dining Room for dinner on the night of December 9, 1968, just that it would be a full house and he'd be expected to make all of them laugh. Johnson was good with a dinner crowd and good at working a room, but he was not especially good at making anyone laugh. Not like Kennedy had been—Kennedy with the wink and the twinkle and all the happy back-and-forth. A press conference with Kennedy had been like an Ivy League cocktail party, all wit and banter. For Johnson it was more like a subcommittee hearing, at least lately, with one question about the war followed by another question about the war followed by a third question about the war, and Johnson looking more and more like a reluctant witness being picked apart by a panel of prosecutors.

But tonight would be Johnson's night, and for once the laughs would come easily. Actually, that was always the way at a White House function when you were president: you tell a joke, you get a laugh. The laughter perk would expire in forty-two days, on January 20, when Richard Nixon, who had defeated Hubert Humphrey in a close presidential election the previous month, would take Johnson's place in the Oval Office.

Johnson was determined to make this evening a happy one, and it was shaping up that way. The dinner was being held in honor of NASA, the Apollo astronauts in particular, and, more specifically, three of those astronauts—Frank Borman, Jim Lovell, and Bill Anders. After the dinner, those three would say good-bye to their wives, head straight back to Cape Kennedy, and move into the on-base crew quarters, where they would live until the morning of December 21, when they would lift off for the moon.

Twenty other astronauts were here tonight, too, all but four of whom had flown in space at least once. To most of the non-NASA guests, these veterans would give off that magical, beyond-the-visible-spectrum aura that only spacemen could. Jim Webb and Wernher von Braun were also in attendance, as was Kurt Debus, the director of the Cape Kennedy launch facility. All three men were integral to the series of programs that had launched the astronauts into space, but they had never gone themselves and thus did not have the aura and never would.

The only other person on the guest list who did have a bit of that special shimmer was Charles Lindbergh, the first man to fly solo across the Atlantic Ocean. Now sixty-six, Lindbergh had taken a keen interest in the Apollo 8 mission and planned to travel to the Cape to watch the launch. Even the spacemen showed a respectful deference in the presence of Lindbergh. The rest of the guests—Vice President Humphrey, cabinet members, senators and representatives—would be mere extras and supernumeraries in an event that would star the aviator and the astronauts.

The evening was to begin with dinner and end with an abbreviated version of the four-act opera *Voyage to the Moon*, written by the German-French composer Jacques Offenbach in 1875. In the twentieth-century version, a young prince takes off in a three-man rocket ship to meet the girl in the moon, even as the king of the moon is hatching plans to push the Earth aside in its orbit so that the moon can get more sunlight. Ultimately there is resolution and romance and peace, and the leaders of the two worlds decide to work together in friendly scientific pursuits. The unsubtle nod to the American-Soviet space race was impossible to miss, though it was open to doubt whether the NASA men in the audience, who had a real moon flight to think about, would be paying enough attention to perceive the intended meaning.

For that reason as much as any, Johnson was determined that his pre-performance remarks succeed, and his writers had given him material that they knew would land well. He began his remarks, and after a couple of easy jokes about how busy the astronauts were and how hard it had been to schedule the dinner, Johnson turned his attention to Webb.

Just before the dinner, Johnson said, Webb "came up to me and said in that serious way of his, 'Mr. President, are you a turtle?'" At this, the astronauts roared, directing their attention now toward Wally Schirra, who practically owned the joke. One of the speechwriters had done some good research and gotten the backstory from the head of the Manned Spacecraft Center's public affairs office.

"This is an in-joke known to the astronauts," read a note on the president's draft. "Unless the person answers with the reply, 'You bet your ass I am,' he has to buy a round of drinks."

The joke may have needed some explaining to the president, but not to the astronauts. And if the other dignitaries in the room didn't get it, that was just as well, since it reinforced the sense that tonight's event was for the fliers' fraternity. Johnson went on in this vein for another minute or two. Then, a politician down to his marrow, he knew when to quit the monkeyshines and get down to the real purpose of the evening. The dinner tonight, he said, was dedicated not just to the Apollo 8 crew but to all of the people of NASA and especially the twenty other astronauts in the room who might one day be making lunar voyages of their own.

"Our rockets can fly from place to place," the president said, "but only the mind of man can cross the new frontiers of space. This may be an unusual toast, because I would like to ask more people to remain seated than to stand. I would like to ask the vice president and the secretary of defense and the members of Congress and their wives to join Mrs. Johnson and me in a toast to the brave and dedicated men of our space program and their wonderful wives."

Chairs scraped as the dignitaries stood and toasted. And although the speech had done what it was supposed to do—pay respects, yes, but also indulge in a bit of the bonhomie of the pilots, the ribbing that would naturally precede a risky mission and help banish the fear that lurked about its edges—there was, all the same, a shadow in the room.

On the list of honored guests tonight—and remaining seated at their own tables to receive the toast with the rest of the NASA family—were Mrs. Gus Grissom and Mrs. Ed White. Unlike the other women in the room, they had come to the dinner unescorted.

☆ ☆ ☆

If there was a great big party taking place in Florida on the weekend of December 20, 1968—and there was—the Apollo 8 astronauts weren't invited to it. Nobody counted the precise number of people pouring out of the cars that were swarming toward the Cape and parking along its beach roads and in its motel lots, but the best guess put it at about a quarter of a million. Likewise, no one could say how far all of these people had traveled to get here, but the license plates came from dozens of states as well as Canada—far more than just Florida's Deep South neighbors, whose residents could usually be counted on to show up no matter how routine the launch.

The people at NASA paid little attention to the horde of spectators camping in the chilly night air the evening before the liftoff. They were much more interested in a more rarefied group: the celebrities and other VIPs who'd been invited as official guests of NASA. Every launch brought a crowd of notables, but this time the list was enormous. At least two thousand people long, it included all of the members of Congress, the Supreme Court, and the cabinet, as well as the aristocracy of Hollywood and industry. Those special guests—who got first claim on the best hotels and sole claim to the comfortable viewing stands on the space center grounds on the banks of the Banana River—had received an engraved invitation to the event.

You are cordially invited
to attend the departure of United States
spaceship Apollo 8 on its voyage around the moon.
Departing from launch complex 39
Kennedy Space Center, with the
launch window commencing at 7 A.M.,
the twenty-first of December,
nineteen hundred and sixty-eight

The phrasing was self-consciously grand and strangely quaint—the business about "spaceship Apollo 8" especially, which sounded equal parts fairy tale and Homer's *Odyssey*. Not a single soul in NASA had ever referred to any of the vehicles they'd ever launched as a spaceship, but the epochal tone of the invitation seemed suited to the occasion.

In the days leading up to liftoff, Borman, Lovell, and Anders saw little of the crowd or the VIPs or the reporters who were swarming around the Cape, and they liked that just fine, especially Borman. The hullabaloo surrounding the mission was a distraction, and not one he welcomed. Borman found the dinner at the White House pleasant enough; it was a nuisance but a tolerable one, and Susan clearly enjoyed it. Shortly afterward, she and Valerie Anders would be returning home to Houston to watch the launch on TV, while Marilyn Lovell and the four Lovell children would go to the Cape. If Susan had to say good-bye to Frank, doing so on the morning after being feted at the White House seemed fair and fitting, given the risks that he—and, by extension, the whole Borman family—was taking.

Once the White House business was done, Borman found the spartan crew quarters on the Cape Kennedy grounds something of a relief. The astronauts were installed in what amounted to a three-bedroom apartment, with a common area, a bathroom, and a small kitchen. The beds were metal-framed and military grade; the dressers and nightstands were slightly better motel grade. The sofa in the common area was a grudging thing, soft enough to be *called* a sofa, but no more.

The one concession to the astronauts' comfort was the cook who had been assigned to them and who was rather grandly called a chef. It was an odd word to have chosen, since the man had actually learned his craft slinging meals for tugboat crews in the Florida ports. That, however, made the food he cooked just right—heavy, simple, satisfying—even if the too-grand title was all wrong.

The ten days at the Cape before the launch would be spent shuttling between the crew quarters and the Apollo simulators, with intermittent briefings by the trajectory experts and flight planners. Still, even in semi-isolation, the crew would suffer the occasional intrusions, which generally came from the odd celebrity who asked NASA for an audience with the astronauts and then was persistent enough to ask more than once.

A few days after the crew settled in, Chuck Deiterich, a mission controller who worked the retrofire console, came to visit. Ordinarily, retrofire was a procedure performed at the end of an Earth orbit mission, when the ship turned itself blunt-end forward and lit its engine to slow down and come home. But on Apollo 8, the spacecraft would execute that same maneuver as it approached the moon in order to enter orbit, then fire the engine again to come home, and then, three days later, have to reenter the Earth's atmosphere from a distance and at a speed that had never been attempted before. If Deiterich had even a small detail to review or address, Borman wanted to hear it.

Today the flight controller and the astronaut were talking about the engine burn required to enter lunar orbit. Sitting in the common area, Deiterich picked up a can of shaving cream that one of the astronauts had left on the coffee table, then turned it bottom forward to illustrate a point about the spacecraft's orientation during the firing. At that moment the two men heard a knock on the door; an apologetic-looking public affairs officer poked his head inside. He was escorting Arthur Godfrey, the redheaded comedian and TV pitchman, who in recent years had been less well known for his comedy work than for his job selling Chesterfield cigarettes, urging his audiences to "Buy 'em by the carton."

Now he came in, walking with a cane. Godfrey had been battling lung cancer since 1959, the wages of the Chesterfields he'd since given up; looking much older than his sixty-five years, he was seeking a moment in the presence of the men who would go to the moon. Borman stood and smiled and shook Godfrey's hand, though the shaving cream can in Deiterich's hand was the only thing that really concerned him at the moment. He accepted the entertainer's best wishes, thanked him for his time, and then the little ritual was over. The public affairs officer nodded a discreet thanks to the astronaut and ushered out the celebrity, who would go home with a "guess who I met" story to tell. Borman, had he been going home at the end of the day, would likely not have done the same.

More important—and more complicated—was the visit by Charles Lindbergh, which came later. Lindbergh had shadowed the astronauts down to the Cape just as he'd said he would, but he didn't come knocking until the day before the launch. Borman, for one, didn't know quite what to make of him. Both he and Lovell had been born in 1928, just one year after Lindbergh's flight, and they'd grown up adoring the great aviator. Then, like almost every American of that era, they'd wound up reviling the man after he made common cause with the German Reich in the late 1930s. Spouting much of the same swill about racial supremacy the Nazis ladled out, Lindbergh had become a leader of the America First movement, arguing that the United States should stay out of the coming European war. In the Borman household, Lindbergh became a man spoken of only with disdain.

Now, three decades later, the once-great aviator was tapping on the door of the crew quarters and asking for a few minutes of the crew's time. The Apollo 8 astronauts let him in around lunchtime on December 20. They all sat down and ate the tugboat man's food, after which the dishes were supposed to be cleared and the visitor was supposed to leave.

Instead, the four men continued talking. Or, more accurately, Lindbergh talked—about the early age of flying, about the aviators he'd known, about the fifty combat missions he'd flown for the United States in the Pacific theater after the war he'd opposed had come anyway and he'd gone to help fight it. He spoke of meeting Robert Goddard, the American inventor of the liquid-fueled rocket, who had told Lindbergh in the 1920s that a trip to the moon might one day be possible but that it could cost an unheard-of $1 million.

The astronauts listened long and attentively. As the afternoon wore on, they at last began to speak about their own mission, which, as a glance at the clock and the lengthening shadows told them, would get under way in less than sixteen hours. At first they spoke almost reluctantly, because if a flight to the moon would be a rather more ambitious affair than a flight across the Atlantic, Lindbergh had already achieved his great deed, while Borman, Lovell, and Anders had yet to do theirs. Lindbergh was attentive as they spoke and asked them a few questions about their spacecraft and their rocket. As they answered, he picked up a piece of notepaper that was sitting on the table and began to scribble something on it, glancing up once or twice to show that he was listening. Finally, he stopped and looked at the three men.

"In the first second of your mission tomorrow," he said, "you will use ten times more fuel than I used on my entire flight." With that, the tarnished old flier showed his respect, making it clear that he would accept no more diffidence from the three young men who were about to make their own historic marks as pilots.

☆ ☆ ☆

Borman and Anders would have a bit of time to reflect on the singular experience of having spent an afternoon with Charles Lindbergh, but Lovell would have no such luxury. While his crewmates had only the imminent mission on their minds, Lovell had his entire family to attend to.

Three days before the launch, Marilyn Lovell had flown to Florida with her two-year-old, Jeffrey, aboard a charter flight arranged by one of the aerospace companies that worked as an Apollo program contractor; such firms invariably scrambled for the goodwill and touch of celebrity that came with giving a lift to an astronaut's family during a launch week. Two days later, the other three Lovell children followed on another charter. The entire family took up residence in a beach house within sight of the launchpad.

Lovell, accordingly, would duck out of the crew quarters when he could, driving to the bungalow to play and roughhouse with his children and then, when they were worn out, walk alone with Marilyn. The two of them did not talk about the mission; there was really nothing to say, just as there had never been anything to say when Jim had been test-flying unproven fighter jets. So they said barely a word about mankind's first flight to the moon, at least not until the night before launch.

Darkness came early on December 20, the shortest day of 1968. At about 5:30 p.m., Lovell drove over to the beach house and bundled Marilyn and all four children into the car. Jim and Marilyn drove in silence, and the children chattered in the backseat as they wheeled onto Highway A1A, toward the space center. After approaching the heavily guarded gate, Lovell flashed his Cape credentials, but the guard did not need to see them, beaming warmly at the lunar astronaut and his handsome family.

More than three miles away, on a spit of land behind the space center fortifications, stood the brilliant spike of white that was the Saturn V. Flood-lit, the rocket was impossible to miss across the flat terrain. Lovell drove through the gate and past the space center's multiple buildings and blockhouses; as he drew closer to the launchpad, the thirty-six-story missile seemed to grow and grow.

The technicians crowding around the pad waved but gave the Lovell family room as they got out of the car on a nearby sand dune and gazed upward, Jeffrey in Marilyn's arms and the rest of the children standing nearby. A NASA protocol group had set up a table on the sand and offered doughnuts and coffee. The Lovells were far too taken with the spectacle of the rocket to think about eating; the technicians, who were used to the sight of a Saturn, were happy for the distraction of a snack. Marilyn had seen the far smaller, steel-gray Titan boosters that had twice before taken her husband to space, but this was different. She craned her neck upward at the Saturn V; if some might find the rocket to be monstrous at such close remove, she found that she had a single, surprising thought:

It's a work of art.

It wasn't a monster at all. It was gorgeous.

Marilyn would watch it fly from a safe distance of one and a half miles the following morning, and it would take the father of her children to a very unsafe distance of nearly a quarter million miles. But at that moment, she could feel nothing but a frisson of excitement for her husband and a deep awe at the machine.

"You know the roar is going to be something terrible," Lovell said as he stepped close to her and followed her gaze.

"We'll be all right," Marilyn assured him. "We saw the Titans go."

"They were nothing like this," Lovell said.

Marilyn nodded.

"And don't worry when it leans," he added.

"Leans?" Marilyn asked.

Lovell nodded. "Just a degree or two to the right at liftoff, so it doesn't hit the tower."

Marilyn tried to picture the massive object tilting even an inch off its perfectly upright line and then shook off the image. If that was the way the thing needed to fly, that was how it would fly.

After a few more minutes, with the children getting restless, Marilyn shooed them back into the car and the family drove off the grounds and returned to the beach house. Lovell joined them inside for a last goodbye, and now, Marilyn noticed, he was carrying a manila envelope. He opened it up and produced a picture of the moon—a close-up of a wide, gray plain, taken by one of NASA's lunar orbiters.

"It's the Sea of Tranquillity," he told her. Then he pointed to a small triangular mountain on the dry bank of the waterless sea. "This is one of the 'initial points' we'll be surveying. It's a landmark that a later crew will use when they begin their descent." Marilyn nodded, not certain why he was sharing this bit of mission arcana. "I'll be one of the first people ever to see it," he said, "so I'm going to name it Mount Marilyn."

Marilyn's eyes filled with tears. Not trusting her voice, she simply gave him a hug and kissed him good-bye.

☆ ☆ ☆

Lovell returned to the astronaut quarters not long after 8:00 p.m., and once he did, the Apollo 8 crew's flight to the moon effectively began. The astronauts were still on the ground, still in their civvies, still breathing the same air and walking the same ground as the other 3.6 billion people on the planet. But their departure clock had begun to run, and it was ticking insistently.

Liftoff would be the next morning at 7:51 a.m. eastern time. That had meant an old man's dinner hour of 5:30 p.m. Lovell had eaten before he went to the beach house; by the time he returned, Borman and Anders were getting ready for bed, and now he did the same. Wake-up, according to NASA's compulsively precise schedule, would come at 2:36 a.m.; a final medical check would follow at 2:51; breakfast would be at 3:21; suit-up would start at 3:56. At 4:42, the crew would walk out to the van that would take them to the launchpad. At 5:03, they would get to the pad, and by 5:11 they would have ridden the gantry elevator to the top and climbed into their spacecraft.

That was how the crew's prelaunch sequence was scripted, and when the planned moments arrived, that was exactly how it began to play out. The reveille came the way it always did, with Deke Slayton letting himself into the astronauts' darkened suite and flipping the lights on in the common room. He then went from bedroom to bedroom, knocking on each door, opening it up so that the light could flood in, pointing at his watch and reminding the crew of what time it was and that medical call was in fifteen minutes.

The menu of steak, eggs, toast, fruit, juice, and coffee was the same as it had been since the Mercury days, and the practice of having the backup crew tuck in with the prime crew remained unchanged, too. This morning, that meant breakfasting with Neil Armstrong, Buzz Aldrin, and rookie Fred Haise. Lovell looked their way, reflecting that but for Michael Collins's bad back, he would have been a member of that trio of stay-at-home astronauts today. Lovell had only recently come to know Haise—the man who'd replaced him on the backup crew when he'd replaced Collins on the prime crew—but he liked what he saw. The newcomer had made himself as much of a LEM expert as Anders was, but unlike Anders, he might actually get to fly the machine.

The reporters massing outside in the predawn chill would not see the astronauts until they were suited up and ready to make the walk to the van. But there was a pool photographer present at the breakfast, and he would follow the crew to the suit-up. Though the astronauts didn't mind the cameraman watching them eat, the suit-up photos were another matter.

A man being dressed for space was a lot less like a knight being dressed for battle than the public would want to know. It was a slow and cumbersome process: each piece of space suit the astronaut donned left him more helpless and thus more dependent on the technicians for every step to follow. They would put on his boots and hoist up his pants and snap the wrist rings of his gloves into place on his sleeves. The result was a puffy, clumsy parade float of a man who, in a final indignity, would be required to spend the next several minutes lying supine on a large recliner—a beached turtle on his back—pre-breathing the canned air in his life-support systems to make sure he adjusted properly. Only then would he be helped back to his feet for the walk out to the van that would take him to the pad.

As the astronauts went through the awkward business of putting on their clothing, the great, clanking machine of Cape Kennedy on a launch day stirred to life around them. The rocket was already fueled and standing on the pad—6.5 million pounds of fuel and machine, plus an additional 1,200 pounds of weight that would be added as the humid Florida air condensed into frost along the skin of the rocket, which had been chilled to freezing by the supercooled liquid oxygen and hydrogen filling the fuel tanks inside.

On the Cape's beaches, the spectators emerged from their tents and cars, blinking into the rising sun and training their binoculars on the rocket more than three miles away. Inside the sprawling firing room, 350 men sat at their consoles, manning a Mission Control far larger than the one in Houston. Yet this operation would be powered up for use only this morning; the moment the engine bells of the rocket cleared the launch tower and Houston took control, the firing room would have nothing at all to do with the flight.

Rocco Petrone, the launch director, surveyed the room from his console at the back, keeping a close eye on both the state of the rocket and the state of his team. A controller who stood at his console, either because he was nervous or simply wanted to stretch his legs, would hear Petrone's voice in his headset telling him to sit back down. A successful launch meant a disciplined firing room.

When Borman, Lovell, and Anders at last emerged from the suit-up building, they walked straight into a storm of flashing cameras and shouted questions from the reporters, who were held behind barricades to keep the crew's short path from the door of the building to the door of the transport van clear. The astronauts carried their portable air-conditioning units in one hand and waved with the other. They saw everything around them through the windscreen of their bubble helmets, and they heard the commotion caused by their departure mostly as muffled noise, like sounds from the surface world heard underwater. Once they climbed into the back of the van and the door was closed behind them, even that sound was stilled.

The drive to the pad was spent mostly in silence; so was the ride up the gantry elevator. As the Florida coastline fell away beneath the astronauts, the frosty, steaming flank of the rocket slipped by. The massive American flag decal and the capital-lettered USA and UNITED STATES— written vertically along the first and second stages—were visible through the ice layer, and the letters appeared in reverse order as the men rose higher.

At the top of the gantry came the walk along the caged gangway of the swing arm to the white room, which surrounded the spacecraft. The hatch of the capsule—which looked so much like the one that had killed Grissom, White, and Chaffee, even if it didn't function like it—stood open, waiting for them. Borman, who would fly in the left-hand seat, climbed inside: first in, last out, like any good commander. Anders, in the right-hand seat, climbed in next. Lovell, in the center and directly under the hatch, would be last, and thus he was left alone briefly on the swing arm.

He looked down at the ground far below and noticed for the first time the hundreds of thousands of people and cars—most with their headlights on in the predawn darkness—gathering to watch the liftoff. He noticed, too, that not a single one of those spectators had been allowed within a mile-and-a-half radius of the rocket. A circular, no-go footprint had been stamped around the massive, violent machine that was the center of so much attention. And the crew of Apollo 8 was perched directly atop that machine.

"Maybe they know something we don't," Lovell muttered to himself—joking, mostly.

After making his way to the open hatch, he peeked inside and saw Borman frowning at the instrument panel. Little Christmas decorations hung in front of each seat.

"What is this?" Borman grumbled, as much to himself as to anyone else, though the question was audible through Lovell's headset.

"Guenter," Lovell responded, an answer Borman knew without having to be told.

Guenter was Guenter Wendt, one of the German engineers who had come over with Wernher von Braun after the war and now worked as white room director and pad leader. Wendt was the last person each astronaut would see before the hatch was sealed, and he loved to surprise the crews with his little pranks and props. Plenty of the astronauts liked them, too—Wally Schirra, especially. But Frank Borman was no Wally Schirra, especially on a mission like this one, so he plucked off the decoration in front of his seat, looked back over his shoulder, and gave Wendt a small and, he hoped, believable smile.

"Thanks, Guenter," he said, handing the nonregulation cargo back to him. Lovell and Anders did the same.

Now another member of the closeout crew appeared at the hatch.

One by one, he stepped hard on each astronaut's shoulder, providing him enough purchase to tighten his seat restraints properly. The job had to be done just so, given the violence with which the crew would be shaken in their seats when the engines lit, then slammed forward during flight when the first stage cut off and dropped away, then slammed back when the second stage lit, and then forward and back again when the second stage gave way to the third. The hatch—the improved hatch, the safe hatch, the one that could be popped open in just a few seconds so that astronauts would never again be incinerated in their seats—was then closed and sealed. It would not be opened again until Borman and his crewmates had gone to the moon and come home.

More than an hour went by as the astronauts and the ground worked through their prelaunch checklists, which so far were being completed without any glitches. The more smoothly they went, the less the likelihood of any holds in the countdown and the sooner the mission would be on its way.

"Apollo Saturn launch control," said Jack King, as the launch finally drew near. King was the broadcast voice of NASA, the man who would narrate the countdown and who could hush even the network anchors in their booths, especially when he offered a humanizing detail he'd picked up by listening to the chatter between the cockpit and the control room.

When the astronauts had climbed into their spacecraft, the sun had not been up, but since then the morning had brightened considerably. "T-minus seven minutes and thirty seconds and counting, and still aiming toward our planned liftoff time," King said. "Jim Lovell reported just a few minutes ago that he could see a blue sky and it looked like the sun is out."

The clock raced downhill toward the six-minute mark and the five-minute mark and then the four-minute mark. At three minutes, the tanks began to pressurize; a powerful churning, glugging sound filled the Apollo spacecraft, a much deeper tone than the one Borman and Lovell had heard in their little Gemini atop their Titan booster three Christmastimes ago. In the cockpit, Borman looked to Lovell, who nodded in recognition of the sound. Lovell then turned to Anders, who had no such sense memory, to offer a reassuring nod.

The last three minutes ticked off. When liftoff finally came, it was every bit as violent as it had been with the first two Saturn V launches.

"Liftoff!" Jack King announced as the five main engines erupted in their controlled firestorm.

"This building is shaking under us!" Walter Cronkite called, once again delighting at the display of raw engineering power. "Our camera platform is shaking. But what a beautiful sight. Man is perhaps on the way to the moon if all continues to go well."

That generic *man*, of course, was in fact three men, and for them the experience of liftoff was something else entirely. They were aboard the beast—within the beast—that was shaking Cronkite's building.

"Liftoff, and the clock is running," Borman called as loudly as he could over the roar. The clock on the instrument panel, which had been still while the clocks on the ground counted down, now began to count up.

"Roger. Clock," said Collins, who had been reassigned as one of the three astronauts who would be working as the capcom for this mission.

"Roll-and-pitch program," Borman said, his voice shaking from the power of the 7.5 million pounds of thrust lifting 6.5 million pounds of machine. As it rose into the sky, the Saturn V began to orient itself, pointing its nose just so for the ride to orbit.

The noise inside the cockpit was like nothing the astronauts' simulator training had remotely been able to reproduce. For at least ten seconds—though to Anders it felt like the better part of a minute—the crewmates had no way to communicate with one another, which meant that each man would effectively be on his own in the event of an emergency. The g-forces were lighter than they'd been on the Titan, just over four compared to the seven or eight Borman and Lovell had endured during the Gemini liftoff. But to Anders, the first-timer, the Saturn V's four g's felt like twice that number.

Anders experienced the brute force of the Saturn V in more than just the g-load. The engines at the bottom of the booster were mounted on gimbals, allowing them to pivot one way or the other to keep the whole stack flying in the proper direction. But such minor motion at the base of the 363-foot spire translated to violent thrashing at the top. Anders felt like a bug on the end of a whip.

The vibration in the cockpit was dramatically more severe than it had been on the Titan. Borman, as commander, had the responsibility of turning the abort handle that would carry the command module and the crew up and away from the Saturn in the event that it went awry or threatened to explode. Mission regulations called for him to keep his gloved hand on the handle at all times, and he was not about to break any rule in the first three minutes of the flight. His fear, however, was that the powerful vibrations of the rocket could cause him to turn the handle by accident, ending an intended lunar mission just a few miles above the Atlantic. But as the rocket streaked into the heavens, Borman kept his hand steady.

At two and half minutes, when the Saturn and its crew were 40 miles above the ground and moving at 5,400 miles per hour, the first stage cut off and dropped away, punching the three astronauts forward into their restraints. When the second stage lit, a second later, they were punched violently back into their seats.

For Anders, this whipsaw meant trouble. A few seconds earlier, he had tried to lift his hand toward the instrument panel and it had felt as if a twenty-pound weight were attached to it. The moment he did succeed in reaching forward was just before the second stage lit; when it did light, it caused his hand to slam back into his helmet visor. The metal wrist ring left a nasty scrape across the unbreakable glass. He cursed himself—the rookie of the flight now had a big rookie mark on his faceplate.

Borman may or may not have seen Anders's mishap; if he did, he exercised the commander's prerogative to ignore a small screwup. "The first stage was smooth and this one is smoother," he announced to the ground.

"Roger, smooth and smoother," Collins answered. "Looks good here."

At the eight-minute mark, the trip suddenly became less smooth, as the Saturn V, now shorn of its first stage, began the vibrational bouncing that had nearly torn Apollo 6 apart.

Borman reported the troubling news. "Picking up a slight pogo here," he said.

"Roger, slight pogo," Collins echoed, both men wondering why von

Braun's newly added shock absorbers weren't doing their job. But seconds later, the helium in the tanks proved itself, working whatever bit of mechanical magic it possessed to settle things back down.

"Pogo's damping out," Borman said.

"Understood," Collins said.

And with that, the Saturn V, which had one perfect and one dreadful flight to its name, did every single thing von Braun had built it to do without another instant's trouble. Its second stage cut off and fell away just when it was supposed to; its third stage lit briefly and shut down again, providing just enough of a kick to carry Apollo 8 to a temporary parking orbit around the Earth. The spacecraft's orbit was too low to sustain for an extended mission, but it was perfectly fine for a crew that would not tarry long. While taking some bearings and checking their systems, the astronauts would make less than two circuits around the Earth before firing up their third-stage engine once more to light out, at last, for the moon.

"Apollo 8, Houston. We have you apogee one-oh-three, perigee ninety-nine," Collins called, reading off the high and low points of the orbit in miles.

"One-oh-three, ninety-nine," Lovell answered crisply.

He settled back into his seat, then spoke to his crewmates: "Okay, we can breathe a little bit more, hear a little bit more, huh?" Around him, the stray dust and occasional bolt left behind by the technicians again floated up into view.

"That was quite a ride, wasn't it?" Borman answered, easing back as well.

"Felt like an old freight train," said Anders.

"It *is* an old freight train, pal," Lovell said, snapping off his gloves and taking off his helmet. "Let's get comfortable. This is going to be a long trip."

On that point, Lovell was exactly right. The moon, at the moment of launch, was 233,707 miles from Earth. At the very peak of their current orbit, the Apollo 8 crew still had 233,604 miles to go.

TEN

December 21, 1968

TECHNICALLY, GENE KRANZ DID NOT NEED TO BE PRESENT AT MISsion Control in Houston on the day Apollo 8 launched. The official manning list—the roster of every controller who would sit at every console for each of the three eight-hour shifts during the six-day flight—did not include Kranz's name anywhere. His odd-on, even-off flight schedule had him busy not with the job of flying Apollo 8 but with planning for Apollos 9, 11, 13, 15, and on down the line for however long the moon ships kept flying. For Apollo 8, the prime seat in Mission Control—the flight director's console—would be filled by a rotating cast consisting of Cliff Charlesworth, Milt Windler, and Glynn Lunney, and they would be more than up to the job without Kranz there to second-guess their work.

But what the manning list said and what Kranz wanted were two different things, and for him, Mission Control was the only possible place to be. Kranz loved everything about the great high-ceilinged control room, with its big board of maps and data filling the front wall like a giant movie screen. He loved it so much that he couldn't imagine walking into the control room unless he felt fully prepared—especially on the days a mission was flying.

Kranz prided himself on his deep, typically dreamless sleep. Most nights, he was like a flesh-and-blood version of the huge engines on the ships he flew: when he shut down, he shut down completely. And also like those engines, when he powered back up there was absolutely no missing the fact, though he sometimes needed a little help.

By his own estimate, Kranz owned forty albums of music by John Philip Sousa, but that count was not necessarily accurate, because Marta and the children kept giving him more every time there was a birthday or some other special occasion that warranted a gift. Almost every morning, Kranz would wake up and put on Sousa—maybe "The Stars and Stripes Forever," maybe "Semper Fidelis," maybe "Hands Across the Sea"—just to get started.

He would drive to work with a portable tape player on the seat, listening to still more Sousa and timing his route so he could hit all the green lights as he sped through the bedroom community of League City on his way to the space center. After he arrived, he could not always say if all of the lights had in fact been green. The truth was, he sometimes couldn't remember seeing any of them because he'd been too busy enjoying his music and anticipating his day.

When he got out of his car in the Mission Control parking lot, Kranz would say good morning to Moody, the parking guard with the gold tooth and the sharp military manner who knew the name of every controller and engineer in the building. Moody would give him a wonderful smile and a crisp hello in response. And then, at last, Kranz would enter his control room.

When Kranz was in the Air Force, he had flown F-86 Sabrejets on patrol missions around the demilitarized zone in Korea. He had liked all but one part of the work. The Sabrejet was a single-seater, which meant he did his work alone, so he had very little experience flying with even one copilot, never mind the close band of brothers who manned an airborne war room like the B-17. He would have liked to have experienced that camaraderie, but to him, working in Mission Control was the next best thing.

Kranz could be dropped into Mission Control at any point in any flight and sense just by looking around the room how far into the shift and into the flight plan the controllers were. He could tell by how full the wastebaskets were, how stale the sandwiches looked. He could tell by whether the pizza was congealed or fresh, by whether the coffee smelled fresh or burned.

If things were going well, most of the men would be tending to business at their own consoles. If there was a problem, a huddle would have formed around the relevant console. One huddle meant one problem. More huddles spelled what could be real trouble.

On the morning Apollo 8 launched, even a novice would have sensed that all was going smoothly. Kranz arrived well before liftoff and took a seat near the back of the room, where he could observe unobtrusively but be available immediately if he was needed. As the countdown clock ticked toward zero, there was the usual bracing; as the Saturn V roared toward Earth orbit, there was the usual tension. Once the rocket arrived there, the controllers would have nearly three hours before the next big milestone: translunar injection, or TLI, the engine burn that would send the astronauts moonward.

TLI was a tricky business, involving lighting up the engine on the fifty-nine-foot third stage of the Saturn V, which was still attached to the spacecraft. When it fired, the engine would accelerate Apollo 8 to the proper speed to put it on a lunar trajectory. The third stage would then be jettisoned and sent into a waste-disposal orbit around the sun. All of this looked simple enough in the equations and simulations, but it had never been tried in manned flight before. Still, if anyone in the room was feeling anxious about attempting the maneuver, they weren't showing it.

A relaxed-looking Mike Collins was working the capcom console. Collins was the right man to have on the shift at the beginning of the flight, since he was more familiar to this particular crew than probably any other astronaut in the corps. Chuck Deiterich was at the retrofire console, and Jerry Bostick was at the flight dynamics, or FIDO, console. They were two more crew favorites and two more good choices, since the TLI burn would be in their hands.

Best of all would be the sight of flight directors Charlesworth, Windler, and Lunney. They were three men doing the same job but coming at it from three different angles. Charlesworth, a physicist by training who had worked in the Army's Pershing missile program, was a man who liked flying machines, understood in a deep and intuitive way how they worked, and would take no nonsense from them. Windler reminded Kranz of himself; he understood the systems and the hardware and the men at the consoles with a field general's thoroughness, but he also felt the romance of the enterprise in a way not everyone at those consoles did.

And then there was Lunney, and Lunney was special. Kranz had worked with him the longest, back since they had both been rookies on Kraft's team. Lunney did not have Kranz's brass or flash, but he might have been the most deliberate, most thorough flight director of the bunch. He had a jeweler's eye for any problems in the systems, and he was dead serious about what the flight directors called chasing nits—making notations in his logbook of any tiny glitch in the performance of any system, glitches that might make no difference at the moment but could, for the next director at the same point in the next mission, be critical.

Kranz looked around at the solid team in the humming room and then turned his attention to the great display screen with the familiar flat map of the Earth and the track of an orbiting spacecraft inscribed around it. Soon enough—shortly after TLI—that map would change. For the first time in history, it would switch from a circular route to a translunar route, with the Earth on the left-hand side and the moon on the right and a spacecraft slowly creeping from one end of the board to the other. And then, in less than three days' time, the map would switch again, this time to an orbital map of the moon.

Something, Kranz felt, was about to shift, something bigger and grander than one space flight or one victory in the Cold War with the Soviets. He looked forward to the exhausting business of being back on the console for all of the flights to come. But today he was happy to have less to do. The change this mission would work on the world would be too beautiful to miss.

* * *

Somewhere between 99 and 103 miles above the Earth, Frank Borman, Jim Lovell, and Bill Anders were not thinking epochal thoughts. They were—for the moment, at least—thinking about not throwing up. Borman and Lovell were hardly strangers to space. Borman had spent fourteen days there; Lovell, the world record holder for time aloft, had logged eighteen. But they had spent all of that time sealed inside a Gemini spacecraft, and its ninety-one cubic feet of habitable volume did not even permit a man to get out of his chair. Yes, they'd been weightless, but they'd known it mostly because the things around them floated, not because they did themselves.

The Apollo capsule was different. It didn't have just a comparatively spacious 218 cubic feet; it had 218 *smart* cubic feet, configured in a way that allowed for maximum mobility even with three men competing for the room. There was enough clearance between their couches and the instrument panel for the astronauts to float over one another comfortably, getting from bulkhead to bulkhead or window to window with ease.

The lower equipment bay beneath the foot of the seats provided still more room. As its name suggested, the bay was partly used for stowage, but it was also where the sextant and navigation console were located, meaning that the bay operated as an entirely separate workstation, distinct from the instrument panel. It was even big enough to provide a little privacy if an astronaut wanted to take a nap or needed to use a waste disposal bag. Those, regrettably, had not improved since Gemini.

Once the crew got into orbit, Lovell was the first to unlatch his seat belt and pop out of his couch. He drifted down to the equipment bay to stow his helmet and promptly felt his head swim and his stomach turn over. Taking hold of a solid projection on the spacecraft wall, he held himself steady.

"Be very careful getting out of your seat," he called back up to Borman and Anders. Anders looked quizzical, but Borman, who had talked with Lovell about the possibility of space sickness in so big a ship, knew exactly what Lovell meant.

"Look straight ahead for a while," Borman told Anders. Then he released his seat belt, too, floated down to join Lovell, and felt the same sickly swoon. The steak-and-eggs breakfast he'd wolfed down in the crew quarters earlier that morning swam before his eyes, and he fought to keep the food down where it belonged. Anders, feeling cautious now, stayed where he was for a few moments before unbuckling as well.

Motion sick or not, in the brief two laps around the Earth the astronauts would have a lot to do before the ship was fit for TLI. Borman's job on this mission was the job of any commander, which meant he was fluent in every single system aboard the spacecraft, could fly the ship alone if he had to, and bore the weight of command as well. He would have the final word on anything and everything that happened over the next six days.

Lovell had been tapped as navigator, and given that Apollo 8 would travel much farther than any other craft in the history of space exploration, that was no small responsibility. While the computer had all of the coordinates it would need for the flight stored on magnetic tape, that mechanical brain still needed a human brain to confirm what the machine thought it knew before the main engine or even a single one of the sixteen smaller thrusters could be fired.

Both the computer and Lovell had the location of thirty-five stars committed to memory. These navigational coordinates were every bit as accurate—and every bit as primitive—as they'd been for the mariners who'd used them centuries ago. Lovell would need that knowledge whenever the ship had to make a maneuver of any kind, and there would be a lot of these operations. He was also the keeper of all of the computer's other commands, especially the firing instructions that would be used when the main engine would make its critical burns. If there was a crew member who would call the lower equipment bay home for much of the six days ahead, it was Lovell.

Anders's role was more improvisational. As the astronaut who was perhaps the world's greatest expert on the workings of the lunar module, he was flying a mission that had absolutely no role for that skill.

Instead, his job as mission photographer would be to use some first-rate cameras to produce images that would improve upon the grainy black-and-white photos beamed back by NASA's robot moon probes, which could never compare with a picture preserved on the halide crystals and gelatin emulsion of a piece of photographic film and then hand-carried home to a lab for development. Anders had been sent aloft with a small arsenal of cameras and film magazines; bereft of his LEM, he was determined not to let a square inch of the lunar surface that passed under his eye go unphotographed.

Until the time to do that work came, he would also oversee the command module's life-support systems. This was straightforward enough work—monitoring oxygen levels, heaters, water flow, and more—though it allowed for little margin of error. The very term "life-support systems" made clear what would happen if those systems failed.

Now, as Anders unbuckled, he surveyed his meters and immediately spotted a problem. The pressure indicator on the glycol system that was supposed to cool the instruments was way too low.

"Boy," he said, partly to himself, "it's way down there. Something's fishy."

Borman popped his head up from the equipment bay. "What's wrong?"

"We're getting glycol-discharge pressure way down," Anders answered.

"You lost the glycol pump?"

"Guess so."

"We'll take a look," Borman said, running the spacecraft's schematics in his head. "Are we in primary?"

If Anders answered yes, that would mean trouble. Anders should have configured his instrument panel properly before the spacecraft left the ground, which included setting the glycol system to primary. If the system was indeed active and already causing problems, the crew could always switch to the secondary system, but it would be troubling if technical problems in the spacecraft were occurring in the first hour of what was supposed to be a 148-hour mission.

"You were . . ." Anders began, then looked again at his instrument panel. "Oh, we're in secondary." He flicked the switch to its proper setting. "It's okay."

Borman drifted over, glanced at the indicator, and smiled as the pressure rose to normal. "It looks like it's picked up here," he said, and floated back down.

The error was harmless, both men knew. All the same, Anders was annoyed at himself. This kind of blunder would never have happened if he'd been flying his LEM.

If Borman had seen no reason to reproach a rookie like Anders in his first hour of space flight, he would soon be not so forgiving of Lovell. None of the three men had yet removed the yellow life vests they'd donned before liftoff, a safety precaution in case they had to abort en route to orbit and ditch in the ocean. Since the space suits were so cumbersome, it had been easy to overlook one more layer of bulk. But now, as Lovell floated past the base of Borman's seat while checking on the navigation panel, his life vest's activation tab caught on a protruding strut. A loud pop and hiss sounded.

"Oh shoot!" Lovell exclaimed as the vest began billowing up on his chest.

"What was that?" asked Borman, who had no clear line of sight to Lovell.

"My life jacket."

Borman laughed. "No kidding?" he asked. "Is it blowing up?"

Then, not wanting to miss the show, Borman maneuvered to where Lovell was floating. His laughter stopped.

Lovell looked comical all right, but Borman instantly understood that the situation was not. Under normal circumstances—on Earth, in other words—it would be easy enough to deflate and stow the life vest. But the vest had been inflated by a small canister of pressurized carbon dioxide, and releasing a big blast of CO_2 in a small, enclosed space with a limited supply of breathable air was not a good idea. The spacecraft was equipped with air scrubbers—square canisters, each the size of a large cookie tin, filled with crystalline lithium hydroxide, which would absorb any CO_2 before it could build up to dangerous levels. But like cigarette filters, the canisters would eventually become saturated and would then have to be swapped out for fresh ones. Dirtying up the first filter on the first day of the mission would be a bad way to begin.

Anders saw the commander's face clouding up and stepped in to help. "Why don't you take it off and give it to me," he said to Lovell. "I'll try to take it apart while you watch the panel."

Lovell nodded his thanks and began struggling out of the vest. "It's hard to get off, too," he grumbled.

Borman saw the discomfiture of both men and rearranged his features to something more agreeable. "Yes," he began, "we . . . we can live with a little CO_2." He knew that this was true, but he damn sure wished they didn't have to. Over the next six days, all three men would have to move around the small cabin a lot more carefully.

It fell to Lovell to figure out a solution, and in short order he came up with an elegant one. Gesturing to Anders to give the vest back to him, he floated over to the closest thing to a proper bathroom the spacecraft had—the urine station in the lower equipment bay. Like the system aboard the Gemini, the station was nothing more than a tube attached to a funnel at one end and a small reservoir in the wall at the other. The urine would flow from the astronaut, through the tube, and into the reservoir. A twist of a knob would then vent the unwanted liquid into space, where it would flash-freeze into Schirra's spangly constellation Urion.

Lovell removed the funnel from the end of the tube, opened the life jacket's valve and fitted it into the funnel's place, and then vented the CO_2 invisibly overboard. When he was done, he folded the vest and stowed it in the storage area. The mission's first serious problem was put away with it.

☆ ☆ ☆

The TLI burn that awaited the astronauts exactly two hours, fifty minutes, and forty seconds into their mission—near the end of their second Earth orbit—would require extraordinary precision. For that reason, neither the astronauts nor the onboard computer would execute it. Instead, the burn would be directed by the room-sized computers at Mission Control.

The Saturn's third-stage engine would fire with a thrust of 225,000 pounds for exactly five minutes and eighteen seconds, accelerating Apollo 8 from its orbital clip of 17,500 miles per hour up to what most newscasters said would be 25,000 miles per hour, which they described as "escape velocity." In fact, they were wrong on both counts.

That nice round 25,000 miles per hour would actually be held to 24,200 miles per hour, and that so-called escape velocity would actually be kept at *near* escape velocity, which was a necessary precaution. The way the physics worked, flying away from the Earth was more like climbing uphill from the Earth, with the ascending power of speed battling the downward pull of gravity. Take off with your speedometer at a full 25,000 and you're going to win that tug-of-war—but if you miscalculate and miss the moon, you'll fly off into space and be adrift forever.

Easing back on the gas just a little—shaving a mere 800 miles per hour off the 25,000—means you will ultimately lose that gravitational battle, so if your trajectory is far off and you miss the moon entirely, you will simply arc over and head back home like a ball thrown high into the air that falls back to the ground. It also means that if you miss the moon by only a little, you'll be moving slowly enough to surrender to the moon's gravity, at which point you'll whip around the far side and get hurled back to Earth. In the first case, you'll make a U-turn in empty space; in the second case, you'll speed around the traffic cone of the moon. Either way, you'll remain on what flight planners call a free-return trajectory—and either way, you'll be heading home.

Until the burn took place, Lovell, for one, was content to claim a few minutes of downtime and drift from window to window while reacquainting himself with the singular view of the Earth slowly turning below his spacecraft. It was a vista he'd seen over the course of 206 orbits the first time he was in space; he would see it for a far briefer spell on this flight, but especially since Apollo had five windows to Gemini's mere two, he wanted to make the most of his opportunity.

"Gee, this is the best flight I ever had," he said with a wink to Borman, a comradely reminder of the grind the Gemini 7 mission had been. He looked out the window again and caught a bright flash as the nighttime side of the Earth gave way to the day. "Here comes the sun," he announced.

"Where?" asked Anders, eager to see as many sunrises and sunsets as possible during their brief stay in the Earth's neighborhood.

"Well, take a look," Lovell said, waving him over to the window. "It crept up on us."

Though Borman stole a few glimpses of his own, he was not partial to the idea that his crew would spend their limited time in orbit sightseeing. But since his fellow astronauts were so busy with the windows, he reckoned that he might as well put them to work there. Given how central lunar mapping was to the mission, a lot would ride on keeping the multipaned windows clear, and that would not be easy. Outgassing from the rubber-sealed frames and microscopic debris that had been knocked loose during liftoff or drifted off in zero g could collect between the glass layers. And any moisture clinging to the exterior windows that had been carried up from Earth would flash-freeze in the vacuum of space.

"Let's give Houston a window-status report as far as contamination," Borman ordered. Before Lovell and Anders could even reply, he began doing just that. "Number one window is clean and has lint on it," he radioed to the ground. "It looks like we're already starting to form bits of frost. Number two window has specks of lint on it."

Lovell, at window three, spotted a streak, tried to wipe it away with his thumb, and frowned when it didn't come clean. "There is a smudge that appears to be on the outside pane," he announced.

Anders reported good visibility out of the center window but added, "There's some dust on both the inside and outside." The word "dust" was more optimistic than accurate, implying a problem that might simply drift away, even though he and everyone else listening knew that this would not be so.

"You're looking good, Apollo 8," Mike Collins said. "We don't have anything for you. We are just standing by."

The remainder of the standby time ticked slowly away. To avoid bothering the crew, Collins held his tongue; instead, he listened to the chatter on Mission Control's internal loop as the other controllers confirmed that the computers and guidance system and third-stage tank pressures were all what they should be. Finally, twenty-three minutes before the critical burn, Charlesworth, who was in the flight director's seat for this first shift, gave Collins the high sign: it was time to signal the astronauts that their ship was fit to leave home. Collins nodded.

"All right, Apollo 8," he said. "You are go for TLI. Over."

"Roger," Borman answered, his voice without inflection. "We understand we are go for TLI. Over."

Collins slumped back in his seat. What should have been one of the most thrilling moments of his life was actually one of the most unsatisfying.

Three human beings, he reflected, were about to tear themselves away from the close gravitational grip of the Earth, and in three days' time they would surrender to the gravity of another celestial body. No living creature had ever done that before. There ought to be an oompah band, he thought. There ought to be fireworks. There ought to be *some* way to mark the moment. Instead, there was just this dull, flat scrap of language: *You are go for TLI.*

But the jargon was deliberate: it was designed to hollow out those very feelings of momentousness, because feelings like those could be distractions when you needed to focus only on the task at hand. Elsewhere in

Mission Control, Jerry Bostick, at the flight dynamics console, allowed himself a flicker of the forbidden wonder.

"They're leaving us," he muttered to no one in particular. "We've got these guys headed out of orbit."

In the spacecraft, Borman and his crew did what the flight plan directed, which was to return to their couches and buckle their restraints. There would be no need for anyone to stand on their shoulders; this time, the belts would simply prevent them from drifting out of their seats. Once the engine was lit, the acceleration would create just enough gravity to settle them back into place.

"All right, gentlemen," Borman said to Lovell and Anders, "let's get set for this."

The astronauts scanned their instruments and saw nothing amiss in alignment or fuel pressure or anything else. Then, just as their training called them to do, they scanned their instruments again. All seemed in order.

At the five-minute mark before TLI, Collins called the crew. "Apollo 8, Houston," he said. "You're looking good down here. Everything looks good."

"Roger, understand," Borman answered.

As those final five minutes melted away, the crews in the spacecraft and on the ground held their silence as best they could. Anders ran the TLI checklist through his mind for the thousandth time, both the systems that were his responsibility and those that weren't. He pictured especially the EMS, or entry monitoring system—the visual readout of speed, trajectory, and spacecraft attitude on Borman's side of the instrument panel.

"You're in EMS, auto?" he asked Borman. The question bordered on impertinent.

Borman nodded yes.

"Apollo 8, Houston, coming up on twenty seconds till ignition," Collins said, his eyes on the clock. "Mark it," he called out. "You're looking good."

"Okay," said Anders.

"Roger," said Borman.

Lovell gave Anders another encouraging smile, then turned to

Borman, who did not glance back. The commander's eyes were fixed on the instrument panel clock and the ignition light next to it.

"Nine, eight, seven . . ." Borman announced, then counted in his head for a couple of digits. "Four, three, two . . ." he resumed.

Behind them, the crew felt a rumble. Fifty-nine feet away, the liquid oxygen and liquid hydrogen flowed from their separate tanks and mixed in the combustion chamber. As the ignition system engaged, the exhaust exploded out of the third-stage engine bell—silent in the vacuum of space, it created a low, vibrating rumble inside the spacecraft.

"Light on," Borman called. "Ignition."

"Roger, ignition," Collins said.

The expected shadow of gravity nudged the crew from behind. Instruments throughout the command and service module, tuned like seismographs to every change in velocity or orientation of the spacecraft, twitched in reaction to the sudden acceleration, turning what they recorded into digital signals and displaying them on the instrument panel. Far more precise sensors—written into the brains and vestibular systems of the astronauts themselves, the result of years and years of flying—reacted, too.

"Boy, it's going off in yaw," Borman said, not liking the slight slewing to the side that both his gut felt and his alignment indicators confirmed.

"It's okay, the DAP is fine over here," Anders said, referring to the digital autopilot.

Lovell was busy minding the other axes in which the spacecraft could drift. "What's your attitude at—" he began.

Anticipating the question, Borman cut him off: "Fine, forty-five . . ."

"Okay," Lovell said.

"And the tank pressure?" Anders asked.

"Tank pressures are good," Borman said.

"Okay," Anders answered.

For five minutes and twenty seconds, the men alternated between silence and necessary chatter. The engine burned, and the ship slowly began to climb that gravity mountain from the Earth to the moon.

Collins called out periodic encouragement.

"Apollo 8, Houston," he said. "You're looking good here, right down the center line."

"Apollo 8," he repeated just over a minute later, "you are looking good. Right down the old center line."

"Roger," Borman answered this time.

This time it was Lovell's job to keep his eyes fixed on the instrument panel clock. He called off the three-minute, two-minute, and one-minute marks to shutdown.

"Thirty seconds to go," he called.

Then: "Ten seconds. Really fine."

Then: "Five, four . . ."

He trailed off.

Finally, three seconds later, just as suddenly as the engine had roared to life, it went completely still. The only sounds that filled the cockpit now were the whirring of the cabin fan, the breathing of the three men, and the crackle in their headsets from the command center. Yet they were leaving the Earth behind at an unheard-of 24,200 miles per hour.

"Okay, we got SECO right on the money," Borman said, as casually as if he were announcing that he'd just picked up the mail.

"Roger," Collins said, "understand SECO."

The moment was marked. Once again, a bit of chilly jargon— meaning "sustainer engine cut-off," or shutdown of the spacecraft's big motor—stood in for whatever emotions the men had been born to feel but had fought and trained not to feel.

In Mission Control, Gene Kranz, exercising the privilege of the spectator, allowed himself to feel plenty. Standing at the back of the room, he watched as the big map on the wall changed: now it showed the long journey ahead. Three men had broken away from the planet. The first mission to the moon had well and truly begun.

ELEVEN

December 21, 1968

VALERIE ANDERS CHOSE TO WATCH THE LAUNCH OF APOLLO 8 IN the way that made her most comfortable—that is, in no special way at all. She would not sit in the VIP stands on the space center grounds, among a crowd of celebrities and the family members of NASA's astronauts. She would not smell the rocket fuel or feel the ground shake or have to shade her eyes to follow the Saturn's rise into the sky. Instead, she would perch on a wooden toy chest in the family den with her youngest child in her lap, the rest of her children at her feet and a television set in front of her. What the moment might lack in drama, it would make up in comfortable familiarity.

This would by no means be the first launch Valerie and her family had watched on TV in their den, but it would be the first they watched on a *color* TV. The Anderses had been among the last families in the neighborhood to trade up to that particular luxury. But given that this would be Bill's first flight and that he'd be away from his family on the very week practically every other family in the country would be gathering together, Bill had decided that he ought to leave a special gift

behind—particularly if it would allow his brood to have a better view of
the beginning of his adventure.

Hours before launch, the Anders home had begun filling up with
astronauts and their families. So had the Borman home, and the Lovell
home would do the same once Marilyn and her children got back from
the Cape. Dodie Hamblin, the *Life* magazine reporter whom the wives
trusted—and therefore NASA trusted—was in suburban Houston today,
and she would do the same kind of reporting she usually did, which
involved visiting the astronauts' homes, doing a lot of quiet observing,
and asking very few questions until the moment was right and the answer
might yield something more than the wifely bromides other reporters
got. There was a reason NASA had made their original deal with *Life*
and a reason they kept renewing it, and Hamblin was a big part of that.

But early that morning, the focus was on the television. Watching a
Saturn V fly on a TV screen may have been nothing like watching it fly
in person, but it was still very dramatic. You could sense its massive size,
and even through the tinny speaker of an ordinary TV, you could hear
its explosive roar. As Valerie watched the five main engines light—in full,
living color—and saw the rocket rise and heard Walter Cronkite shout
about his booth shaking and man perhaps being on the way to the moon,
she had only one thought: "Thank you, Dr. von Braun."

It was odd maybe, but it was apt. He had built the rocket, and that
rocket, she could already tell, was flying true. She stayed to watch the
Saturn do its job, shedding its first stage and dwindling to a dot. She
watched as the TV animations took over to show the parts of the mission the cameras no longer could. She heard the reporters announce that
her husband was in orbit around the Earth. And then, less than three
hours later, she learned that he had left that orbit and was headed for
the moon.

Valerie knew that she would need to take the mission as a series of
such moments and milestones and that she'd have to pace herself
emotionally—and pace her children, too. This was especially important
since they would have to spend so much of the next six days inside the
house.

The day before, she had planned to drive over to the space center
commissary to lay in some supplies for what she knew would be a week

of siege. A few journalists had already arrived and taken up positions in front of her house; by launch morning, the crowd of reporters and photographers would grow large enough to encircle the house, effectively imprisoning her family inside. She'd known that she had to get out early or not go at all. There were supermarkets in town, but she would surely draw notice there. At the space center, where people were forever running into an astronaut or one of the other wives, she would be nothing special.

To avoid the early-arriving journalists, Valerie had tried to slip out of the house through the children's playroom door, which opened onto the backyard and was concealed by a wooden fence surrounding the property. Carrying Eric, the youngest, who at four years old would still suck his thumb when he was tired or overwhelmed—both of which he was today—she worked the gate open with her free hand. As soon as she did, she walked straight into a photographer. He snapped a picture of an ambushed mother and a startled-looking toddler who had just popped his thumb out of his mouth.

"Thumbs up for Dad!" the caption under the photograph in the next day's paper would read.

Valerie, defeated even in that small venture outside, retreated indoors to begin her confinement early. Before long, however, she realized that she needn't have worried about going shopping. By that evening, her refrigerator was stuffed with casseroles and sandwiches and potato salad and snacks, and her counters were stacked with pies and coffee cakes and cookies, all provided by the visitors who had been through this experience before and would never think of coming to the home of an astronaut on a mission without provisions.

If there was a measure of privacy to be had, it came from an accommodation NASA always made for the astronauts' families: a small squawk box that was typically set up in a bedroom or some other less public spot. From the moment the crew took off until the moment they splashed down, the device allowed the family to listen to every second of the air-to-ground chatter. At first, the transmission was delayed by a fraction of a second; later, when the spacecraft got to the moon, that would grow to a full second and a half—the lag representing the time it took a radio signal, even traveling at 186,000 miles per second, the speed of light, to cover the 233,000-mile translunar distance. In truth, NASA had built an extra couple of seconds into the transmission, allowing for a kill switch to spare the families from hearing something they shouldn't. Nobody in Mission Control who had heard the voices crying out from the burning Apollo 1 spacecraft could abide the idea of that horror pouring straight into the ears of the dying man's wife or child.

Before Bill left, he had been honest with Valerie about the risks he faced and the odds that he would come home. "There's a thirty-three percent chance the mission is a success, a thirty-three percent chance we come back safely but don't make it to the moon, and a thirty-three percent chance we don't come back at all," he'd said. Anders's odds-making sense came from the same place as Chris Kraft's less rosy fifty-fifty prediction: his gut. And although neither man could say with certainty how he had arrived at these figures, both were confident that they were right.

Planning for every possibility, Bill had left behind two tape recordings for the children. They were to listen to the first tape on Christmas Day; they were to listen to the second tape only if it became clear that the family would never again spend Christmas together.

Valerie gave as little thought as possible to that second tape. As a pilot's wife, she had long since developed a kill switch of her own, one that allowed her to consider only the things she could control and ignore the terrible things she couldn't. Many of her friends were pilots' wives, too, and a number of them had husbands in Vietnam—a very different, far bloodier battle in the Cold War than a mission to the moon. If they could shut out the fear, she thought, she could do it, too. What's more, she would help her children learn the trick as well.

She knew she could not promise her children that their father would come home, and therefore she wouldn't. But she could promise them that they had her, and that they always would.

"I'm here," she told them as she tucked them into bed the night after the rocket was launched. "And I'll be here." She planned to repeat that every night until Bill was back.

☆ ☆ ☆

Susan Borman's sons did not need to be reassured the way Valerie Anders's children did—or if they did need it, they weren't about to let on. Fred was seventeen and Ed was fifteen, and both were now taller than their five-foot-seven-inch father. They were tall enough to be on the high school football team, in fact, and tough enough for it, too. And they were old enough to believe that there were emotions a man displayed and emotions he didn't, and they guessed they knew the difference.

The boys got their innate calm from their father, but they got another kind—the kind you could put on even if you weren't feeling it—from their mother. There were days when Susan needed that skill more than others, and Apollo 8's launch day was one of them. Friends and family began arriving before dawn, and she stayed busy entertaining her guests, keeping watch over her sons, and sparing them the peckings of the press. There were no reporters in the house during liftoff, but after Apollo 8 was safely in space, she gave the gathering journalists what they needed— on her terms—emerging on the front lawn with Frank's parents and the family dog, and without Fred and Ed. After the cameras finished capturing that tableau, she offered up a few words: "I'm always known as the person who had something to say, but today I'm speechless."

But then she did go on to speak, answering the predictable questions with the proper declarations of confidence and pride. "The magnitude of this entire thing is very difficult to comprehend and hasn't sunk in on me. This is very much different from Gemini 7."

Finally, toward the end of the brief session, she said, "I'm too emotionally drained to talk."

Then, begging the press's pardon, she went inside. She had a very long week ahead and knew by now exactly how much attention she could spare for the reporters on any given day. For launch day, she had no more to give them.

☆ ☆ ☆

Before the astronauts of Apollo 8 had even shed their heavy pressure suits and donned the white jumpsuits they would wear throughout their mission, they had already traveled farther from Earth than any person ever had. For more than two years, Pete Conrad and Dick Gordon had held the altitude record for humans, having flown their Gemini 11 spacecraft to a then-unprecedented 850 miles up. It took a very long time for the species to reach that pinnacle, but as Borman, Lovell, and Anders sped toward the moon at their near-escape velocity, they put that kind of mileage on the odometer every ninety-one seconds. By the time they reached the thirty-five-minute mark after the TLI burn, they had beaten Gemini 11 more than fourteen-fold, climbing to nearly 14,000 miles.

Lovell, who had flown with Conrad at Naval Air Station Patuxent River when they were both test pilots, delighted in needling his long-time friend. "Tell Conrad he lost his record," he radioed down with a smile.

Apollo 8's spectacular altitude should have meant an equally spectacular view of the Earth, but the crew couldn't yet think about turning the spacecraft around to get a good look at their home planet. Soon enough they would be able to do that; once the ship was headed in one direction, physics dictated that it would continue moving in that direction no matter which way its nose was pointed. For the moment, though, the more important issue was that the Saturn V's third stage was still hanging off the back of their service module. The rocket's final stage had done its job well, but now it was space junk and it had to go.

The third stage was connected to the spacecraft by a ring of explosive bolts. The separation maneuver called for the crew to detonate the bolts, then pulse their thrusters to add a few extra feet per second to their speed. That would open a gap between the spacecraft and the third stage. And the gap couldn't be small, either: when the bolts blew, the stage would begin leaking stray fuel through the severed lines, making its behavior unpredictable—the last quality you want in a twenty-three-thousand-pound piece of hardware that's trailing right behind you. Borman preferred to avoid that kind of randomness.

Each of the astronauts had a part to play in the delicate maneuver. Anders would have the flight plan checklist in hand and read off the separation commands; Lovell would punch them into the computer and execute the explosive separation; Borman would control the thrusters once the spacecraft was free. Lovell, for one, had been looking forward to this first opportunity to operate the computer in flight: he had spent many hours practicing with the powerful electronic brain on the ground, and now it was finally time to test it for real.

The computer, which was operated via a nine-button keypad, was relatively compact, but it had a very large screen that was able to display twenty-one characters in a single line from left to right. The language of the machine was, in its way, a great deal like spoken English, consisting principally of verbs and nouns that were represented by numbers. During months of training, Lovell had made it his business to learn how to speak the computer's language fluently. A verb represented some action that was to be taken, and a noun represented the thing that was supposed to be acted upon. Punch in the verb 82, which stood for "request orbital parameter display," and the computer would digest the command and then wait for more. Which orbital parameters exactly? Inclination? Velocity? There were a lot of them. Following the 82 with a 43—for latitude, longitude, and altitude—would complete the command, and the computer would respond. Fortunately, the complex separation maneuver would be made somewhat easier because part of the procedure was preloaded, which meant the computer had all of the nouns memorized. All Lovell needed to provide were the verbs.

Borman scanned his instrument display to make sure the ship was configured properly for the maneuver. It was.

"All right," he said with a nod at his crew.

"Okay, verb 62, enter," Anders read out.

"Verb 62, enter," Lovell confirmed and punched the proper key.

"Verb 49, enter," Anders said.

"Verb 49, enter," Lovell repeated.

Anders scanned his instruments and nodded in approval. "Okay," he said, "proceed."

"Roger," Lovell answered and pressed the button on the display panel that read just that: PROCEED.

The bump that occurred when the bolts exploded took all three men by surprise. It was certainly tolerable, especially compared to the earthquake of the launch, but it was much more of a punch in the back than the simulations had led them to believe it would be.

Borman shook off the jolt, grabbed the pistol-grip handle that fired the spacecraft's thrusters, and began edging forward. Once he had opened up what the instruments told him was a sufficient gap, he planned to pitch the ship backward so that the crew could look through the windows and confirm that distance. But the windows were small and the third stage could be anywhere; it would not be easy to bring it into frame.

Borman fired his jets, performed the half somersault, and looked out his window. Nothing. He nosed left, then right; up, then down. Still nothing.

"Man, where's that S-IVB?" he said. "Anybody see it now?"

Lovell and Anders squinted through their own windows, staying silent as Borman continued to ply the thrusters.

After another moment, Lovell called out: "There it is!"

"You found it?" Borman asked.

"Right in the middle! Right in the middle of my window!"

The third stage was there all right—bright white and reflecting the sun. By eyeball reckoning, it appeared to be as far behind their spacecraft as the computer said it was, which was several dozen yards, but that was not yet far enough. Borman could see that it was spraying so much fuel that it was in danger of tumbling out of control and presenting a collision hazard. Not caring for that possibility, Borman began contemplating whether an evasive maneuver would be necessary.

Then, suddenly, all thoughts of the troublesome third stage fell away, because in that moment he saw something much, much grander. He saw the Earth.

It was a view that American astronauts and Soviet cosmonauts had seen from space many times before, but in those cases, the planet had been a broad arc, too big to fit into the aperture of a window because it was too close. Now, however, Borman, Lovell, and Anders could see the planet floating alone, unsupported, in space. The Earth was no longer the soil beneath their feet or the horizon below their spacecraft. It was an almost complete disk of light suspended in front of them, a delicate Christmas tree ornament made of swirls of blue and white glass. It looked impossibly beautiful—and impossibly breakable.

What Borman said aloud was: "What a view!"

What Borman thought was: *This must be what God sees.*

Then he collected himself. "We see the Earth now, almost as a disk," he radioed down.

"Good show," Collins said. "Get a picture of it."

Borman gestured to Anders, but the prompt wasn't necessary; the mission photographer was already assembling his cameras. Lovell looked out the window and described the scene Anders would capture.

"We have a beautiful view of Florida now," he said. "We can see the Cape, just the point. At the same time, we can see Africa. West Africa is beautiful." Then, to stress the magnitude of what he had just said—the perspective he had—he added, "I can also see Gibraltar *at the same time* I'm looking at Florida." More than that, he could see Cuba, Central America, and most of South America. "All the way down through Argentina and down through Chile," he said.

Borman allowed himself to take in the view for a moment longer, then turned his mind from sightseeing back to business. No matter how NASA had expected the third stage to behave at this point in the mission, it clearly wasn't playing along. It was supposed to leak only a little fuel and then conduct what was known as a blow-down maneuver, emptying whatever remained in its tanks in a quick and tidy blast to prevent uncontrolled venting. But Borman was not looking at anything like quick and tidy now.

"Boy, it's pretty spectacular," the commander said. "It's spewing down from all sides like a huge water sprinkler."

"Get some pictures of it," Collins instructed again.

Borman wasn't thrilled with this directive. Photographing the third stage wasn't nearly as important as being done with it; in his mind, this was exactly the kind of freelancing that could end in misery. But the ground wanted pictures, and at least at this point, Borman wasn't about to disobey.

Lovell heard Collins's request, and since the third stage was visible in his window, he took the camera from Anders.

"Could you pitch up a little?" he asked Borman.

The commander complied.

"Could you pitch a little more?"

Borman gave the thrusters another tweak.

Anders moved up to Lovell's window and squinted outside. "We haven't got it in here yet," he said. "Could you pitch just a little more?"

"I'm not going to fly around the damn thing," Borman snapped. He puffed the thrusters a bit more, until Lovell at last had the view he needed and fired off a fusillade of shots.

Anders, glancing at Borman and guessing that his patience was just about worn through, turned back to Lovell. "Don't you think that's enough pictures of it?" he said.

Lovell lowered the camera and handed it back to Anders.

Borman directed one more glower at the misbehaving third stage. Then he looked back down at his thruster pistol grip.

"Houston, Apollo 8," he radioed down. "I suggest a separation maneuver if that's all right with you."

There was silence on the ground. Borman could practically see the flight controllers contemplating this idea, conferring with one another about whether to green-light so routine a procedure as an unplanned thruster firing. He gave them about twenty seconds to consider the matter.

"Houston, Apollo 8," he repeated.

His words were again greeted with silence. This time he gave them just six seconds.

"Roger," he said with finality. "I believe we're going to have to thrust to get away from this thing."

There would be no Wally-type mutiny on a mission Borman commanded, but there would be no dithering, either. He hit his thrusters and Apollo 8 jumped quickly away from the third stage, leaving it to fall into its trash-can orbit around the sun.

☆ ☆ ☆

If the global audience that was following the mission during its first twenty-four hours had been listening carefully, at least a few of its members might have noticed a troubling bit of chatter going on between the spacecraft and the ground. It's a safe bet that no one actually noticed; there was too much else to pay attention to as the spacecraft blew through its 14,000-mile altitude record, increasing it more than eight-fold to 120,000 miles in the time it took the Earth to turn just once.

Even more exciting was the promise of several live TV broadcasts from space. The first was due to occur at the thirty-one-hour point in the mission, or about 3:00 p.m. eastern time on December 22; at that time, the folks at home would be able to see the planet they inhabited from the same surreal perspective enjoyed by the astronauts.

Still, there was that occasional troubling chatter. To the uninformed listener, it would almost certainly be merely puzzling.

Anders, for instance, would say: "Houston, we've rewound the tape. You can dump it at your convenience."

And Collins would respond: "Apollo 8, Houston. We're going to try to dump your tape right now."

A while later, one of the astronauts would say that NASA might enjoy some interesting details on those tapes, and then he would request that NASA let the crew know their opinion of the tapes once they'd had a chance to listen to them.

What the crew and the ground were talking about was the DSE, or data storage equipment, which they more casually referred to as the dump tapes. Installed in Apollo 8's cockpit was a recording system that remained on, more or less continuously, from liftoff to splashdown. The dump tapes recorded everything the astronauts said to one another in the privacy of their spacecraft, without the air-to-ground loop picking up what was being discussed. Not only did the tapes create an important historical record, they would also provide critical information if there were ever an accident on board and a later investigative committee had to determine the cause of the problem.

Most important was that the taping system gave the crew a way to record a private message to Mission Control and then transmit it at high speed without anyone in the larger world listening in. But Mission Control would need to find a spare moment to listen to the damn thing, and at present they were taking their time about doing so. That was a problem, because what the crew wanted the ground to know was that Borman was sick—and it wasn't just the mild motion sickness that Lovell and Anders had experienced early on.

For the better part of twelve hours, Borman had been alternating between throwing up and battling the urge to throw up, a fight he often lost. He was also experiencing intermittent episodes of very loose bowels, a symptom that often accompanies this kind of digestive upset. Both problems were extremely difficult to manage in a spacecraft that had no indoor plumbing. Although the commander was going about his work and his voice betrayed nothing, he would be able to carry on that way for only so long. If Borman couldn't eat food and hold it down, his performance would falter, and eventually he wouldn't be able to function at all. Already, the sound and smell of his suffering were making the cramped cockpit unbearable for all three men. Worse, if Borman's sickness had been caused by a virus, Lovell and Anders would almost certainly contract it, too.

At first, Borman forbade the crew to breathe a word to the ground. "I'm not going to say anything at all," he told them when Houston wasn't listening. "And you guys shut up, too."

But after half a day, Borman himself was worried. He had never gotten sick in an airplane in his life unless he was hungover—which happened at least a few times in any Air Force man's career. During the Gemini 7 mission, his stomach had been rock steady for fourteen straight days. This time, however, his digestive system was in full revolt.

Lovell and Anders knew that if he had to, Borman would just grind it out, and that worried them. But finally they persuaded him to sign off on the dump-tape plan, which would eventually bring the problem to the attention of the flight surgeon, Dr. Charles Berry—or so they hoped.

Hours after the recorded transmissions had been sent, Houston at last picked up the hints the crew was sending.

Glynn Lunney was at the flight director's console; Cliff Charlesworth, though he was not on duty, was nearby. They both suspected that whatever the crew was going on about, it might be a medical issue. Lunney asked Charlesworth to summon Berry and said that the two of them should meet him in a backup control room just one floor down; there they could listen to the tapes in private and communicate with the crew if they needed to. Berry, in turn, called Collins, who had recently ended his shift on the capcom console and turned the microphone over to rookie astronaut Ken Mattingly.

The four men arrived in the backup room, shut the door, and listened to the tapes in mounting alarm. The best case was motion sickness. The not-good case was a virus. The worst case—the one that occurred to Berry immediately—was radiation sickness. It was consistent with the sudden onset; it was consistent with the vomiting; it was consistent with the diarrhea. And on this mission, a ready source of radiation poisoning had been impossible to avoid: the Van Allen belts, the bands of radiation that surround the Earth from a low of 620 miles up to a high of 37,000 miles.

Gemini 11 had briefly grazed the lower edge of the belts. But the Apollo 8 crew had plowed right through them, getting a full dose of the high-energy rays, with little to stop them but the comparatively thin skin of the spacecraft itself. Even at the high speed the ship was traveling, the crew would need two hours to clear the upper limits of the radiation field.

Berry had been fretting about the risks the Van Allen belts posed since the beginning of the Apollo program, but there was no way off the planet Earth except to push through those thousands of miles of radiation, so they had built the most robust ship they could and hoped the crew would not suffer any ill effects. Now Borman was displaying exactly the right symptoms at exactly the right point in time.

Berry raised the radiation-poisoning possibility to Lunney, Charlesworth, and Collins, but they weren't persuaded. Lovell and Anders were healthy, weren't they?

So far, Berry answered.

And even if the belt extended more than 36,000 miles high, they reminded him, the radiation dose was low all the way through—little more than what a man would get from a chest X-ray.

Yes, Berry replied, a two-hour chest X-ray.

Still, Berry was a scientist, and he had to agree that if the other two astronauts were not experiencing Borman's symptoms, it was unlikely that radiation sickness was causing the commander's distress.

The likeliest explanation, to his way of thinking, was a virus, and that would create a serious situation. According to flight rules, that left him only one option.

"I'm recommending that we consider canceling the mission," he said.

Lunney, Charlesworth, and Collins looked at him in disbelief. But also according to flight rules, their only option was to get on the radio and send the medical man's opinion up to the ship. Still in the backup control room, they called the crew and got right to the point.

"Dr. Berry thinks you've caught a bug, and he's worried Bill and Jim are going to get it, too," Collins said to Borman. "He's recommending that we consider canceling the mission."

"*What?*" Borman exclaimed. He turned to Lovell and Anders,

equal parts amused and outraged by the suggestion. Looking at his crew-mates and lowering his voice to a mutter, he said, "That is pure, unadul-terated horseshit." The other two men nodded in agreement.

For Houston's benefit, Borman collected himself. "Look, you've got three mature people in a spacecraft here, and we're not just going to turn around and come home. I'm fine." That was hardly the case, but he then amended his fib with the truth: "Or at least I'm feeling better."

And, in fact, he was. The half day it had taken him to admit the problem and the additional half day it had taken NASA to respond to the dump tapes had given his stomach time to settle down. Now Borman was sure the problem wasn't radiation, and he suspected that it wasn't a virus. That left the most ignoble explanation of all: an extreme case of motion sickness. He was the first American astronaut ever to report it, but in the roomy Apollo, he figured, he wouldn't be the last. Either way, he would speak of it no more.

☆ ☆ ☆

The reception was terrible on the one television in the office inside the Central Research Institute building in Moscow. There were other chan-nels to watch—much clearer channels—but they were all state-controlled. If you wanted to get Eurovision, which was what the men in the office needed today, you had to rig a special cable. That was certainly some-thing the engineers who worked in a place like the Central Research Institute building could manage, but it didn't mean the picture would be terribly good.

Still, that was the only way the Soviet space brass could follow the activities of the three men aboard the American spacecraft that was now, just thirty-one hours after its launch, halfway to the moon and about to beam a television show to Earth. Nobody outside the room was officially told the names of all of the people who were inside, but Dmitry Ustinov, the Central Committee member who was being groomed as minister of defense, was certainly among them. So was Victor Litvinov, a gifted air-craft designer who directed the country's aerospace industry. As was Boris Chertok, perhaps the nation's greatest rocket engineer after Sergei Korolev—the late chief designer himself.

Those three and many more officials had watched the Americans take off the day before on a giant viewing screen in Building 88 of the nearby Scientific Research Institute in Moscow, and the sight of the bright white Saturn V had been discouraging enough. The Soviet Union's own N1 rocket—the military-green, heavy-lift vehicle that was their answer to the Saturn—had not yet had a successful unmanned flight, much less a manned one, and here were the Americans trusting the lives of three astronauts to their cursed Saturn.

There had been one moment during the launch that had caused the watching Soviets to catch their breath: when the second-stage engines lit, they produced a huge white cloud that made it appear as if the Saturn had exploded. If there was a man in the room who experienced a flicker of disappointment that it hadn't, he observed the missileman's credo of not giving voice to such a forbidden sentiment. But within seconds it was clear that the Americans were on their way to Earth orbit and then toward the moon, moving one step closer to eliminating the Soviets from the decade-long space race. Whatever their private thoughts, every man who had watched the launch left the room dispirited.

The television show from inside the spacecraft would be even worse, with the happy, cocky Americans showing off for their countrymen back home—and showing up the people of Russia. For that reason, there would be no more big-screen viewings in Building 88. Ustinov and a few others of high rank would be free to watch what they chose in their locked office with its rigging of cable. For everyone else, the TVs would stay dark.

When the broadcast began, the sloppiness of the whole affair was impossible for the Soviets to miss. The three Americans might have been on the way to the moon, but they didn't appear to be taking the job very seriously or doing it very well. The astronaut they called Anders was holding the camera, and he seemed not quite able to make the thing work. He tried to show the Earth out the window, but the exposure was all wrong and the image came through as a washed-out circle.

"We are having no joy," said a voice from the space control center in Houston, in apparent criticism of the astronaut's work.

"How about now?" Anders asked.

"Still no joy. It's coming through as a real bright blob. Hard to see what we're looking at."

The camera then swung around to show the inside of the ship, and the picture became much better. The commander known as Borman appeared on the screen, but he was upside down.

"You have everybody standing on their heads down here," said the voice from Houston again.

"Well, we all have our problems," answered Anders. Then he turned the camera and righted the picture.

The commander began talking. "I certainly wish that I could show you the Earth," Borman said. "It is a beautiful blue with predominantly blue background and just huge covers of white clouds."

The camera moved again and showed the third man, Lovell, in what looked like a storage area beneath the seats, working with a plastic packet of some kind. "Jim, what are you doing here?" said the commander. Answering his own question, Borman said, "Jim is fixing dessert. He is fixing up a bag of chocolate pudding. You can see it kind of floating by."

Lovell smiled and looked toward the camera, already showing a growth of beard after only a day and a half without a shave. The commander noticed that and said, "Let everybody see that he has already outdistanced us in the beard race. Jim has quite a beard already."

Anders appeared on the screen, and as he handed the camera off to Borman, he picked something up.

"You can see Bill has his toothbrush here," Borman said. "He has been brushing regularly." Then Anders began playing with the toothbrush, letting it go and catching it in the air. "To demonstrate how things float around in zero g," Borman explained. "It looks like he plays for the Astros, the way he tries to catch those things." Presumably, the Soviets figured, this was a reference to an American sports team.

The men went on like that in their happy way for five minutes or so before ending the transmission. "We will be signing off and we will be looking forward to seeing you all again shortly," said Borman.

"Roger," said the voice from the control center.

"Good-bye from Apollo 8," said the commander.

The transmission from space ended, both for audiences watching it in the free world and for anyone who'd managed to get a signal somewhere else. The next day there would be a much larger meeting in the Central Research Institute building—the little television show had made that certain. Ustinov would chair it, and he would ask a question that would surprise no one.

"How are we going to respond to the Americans?" he would demand. "Sort it out and tell me what you're going to do."

Nobody attending the meeting was likely to have a very good answer.

TWELVE

December 22, 1968

ON THE SAME DAY THE MEN IN MOSCOW WERE DISCONSOLATELY watching the broadcast from Apollo 8, others in the Soviet bloc were also turning their eyes toward the mission. The little show from deep space might not have been carried anywhere in the USSR or Eastern Europe, but official blackouts were only so effective, and plenty of people living within that vast sprawl of territory knew that with a little artful rigging of a powerful antenna, they could pluck a forbidden signal straight out of the air. There were not likely to be many broadcast pirates, but there were enough that reports spread quickly about what they had seen.

In the West—and pretty much every other place on the planet that could pick up a television transmission—there was no need for such signal theft, and the show from the spacecraft was widely available. Network executives had long known that despite such obstacles to viewing as lack of access to television sets or even electricity in some parts of the world, as well as official bans like the one in the Soviet Union and certainly in Red China, a show that was big enough or important enough could still be watched by hundreds of millions of people. Though the first broadcast from Apollo 8 almost certainly did not break any records, the networks concluded that the global audience exceeded one hundred million viewers, and it was possibly two or three times that.

People watched the five-minute show in Britain, where the *Sunday Times* put aside its coverage of the country's persistent economic woes and admitted, "It's hard to write about the pound when the Apollo flight is thrilling us all." They watched it in France, where the *Journal du Dimanche* called the flight "the most fantastic story in human history." They watched it in Hong Kong, where broadcasts from Britain's BBC were readily available, even if mainland China was ignoring the mission. "Hong Kong Man on Way to Moon," read the headline in more than one newspaper, the editors delighting in Anders's geographic birthplace while overlooking the fact that a child born on an American military base was considered to have been legally born on American soil.

Hours before the broadcast, Pope Paul VI, in his weekly address to the crowds at the Vatican, had Apollo 8 on his mind, too. "We accompany with our prayers the courageous astronauts, flying in space at a dizzying speed, wishing a happy success to a risky interplanetary voyage," he said.

In the United States itself, all three television networks carried the show, and they fully intended to air every minute of the five other broadcasts from space that NASA had arranged. And why wouldn't they? The ghostly signal showing three ordinary men with ordinary names like Frank and Jim and Bill making an extraordinary journey was an entirely different kind of news from the full-color, close-up bloodshed that had been filling TV screens all year. It was an unalloyed good thing, and the uplift of the moment seemed somehow to be part of a small, rising air current that drew other happy stories along with it.

It was surely just coincidence that on the same day the astronauts beamed their television show home, the eighty-two surviving men of the naval vessel *Pueblo*, who had been held as prisoners in North Korea since January, were at last released by their captors.

"Wonderful!" Frank Borman exclaimed from his spacecraft when Jerry Carr delivered the happy news to him.

And it was surely just coincidence that that same Sunday, Julie Nixon—the pretty twenty-year-old daughter of the incoming president— was married to David Eisenhower, the twenty-year-old grandson of former president Dwight Eisenhower. The entire event had a certain royal family pageantry to it, and if royalty wasn't exactly the kind of thing the American citizenry was supposed to care about, in this unhappy year there was a refreshing gentility to it.

The lightening of the mood seemed to touch president-elect Nixon, too, a man not known for warmth or sentimentality. "I have the honor of giving the second toast of the evening," he said when it was his turn to speak at the wedding reception. "This is by way of breaking me in for the many toasts I will be giving in the future. But none will give me as much pleasure as this. This is also a very newsworthy and happy day. Today I saw the Apollo men halfway to the moon and the crew of the *Pueblo* released."

Nixon may have benefited from a lucky news break on a day that inevitably would have allowed him to showcase his sunnier side, but he was gracious and genuine-sounding all the same. In the shimmer of all of the happy developments, the nation, for once, seemed to be catching a few lucky breaks, too. But whether the run of luck would hold for the three men in the small pod hurtling through the deep void would not be known for a few more days.

☆ ☆ ☆

As far as the people in NASA's public affairs office were concerned, there was entirely too much conversation about balls and urine going on between the Apollo 8 astronauts and Mission Control. The agency's image makers and television experts had feared that this would happen; they'd even discussed the problem among themselves. They all agreed that something ought to be done, but nobody could figure out what, so the talk about balls and urine kept right on going.

The balls part, at least, was less frequent than it had been on earlier missions—and when you were dealing with military aviators like astronauts, that counted as a victory. Ship-to-shore communications were always a scratchy business, which was why call signs or vectors that included letters like *A*, *C*, and *T* became Alpha, Charlie, and Tango for clarity. Nothing really accounted for turning *W* into Whiskey, since the letter had no rhyme that could cause confusion, but Whiskey sounded fun and manly, so it was added to the mix, too.

The same thinking applied to the use of balls. Numbers had no rhymes that could cause problems, so it was fine to call out one, two, and three by their proper names. But zero, which was even less likely to create rhyming difficulties, proved to be an irresistible target, and so aviators referred to zeros as "balls."

Wally Schirra, more than most astronauts, could barely contain himself. Throughout the eleven days of the Apollo 7 mission, Wally had taken special pleasure in calling down, "First off, we'll read off balls," or "Star difference angle was four balls," or "Two balls twenty-two, plus four balls six, plus four balls one."

Inevitably, the capcom would follow that lead, since he could hardly say zero when Wally was talking balls. So the voice from the ground would answer the commander: "Okay, all balls minus twenty-six eighty-seven."

Then, finally, a female reporter at a NASA press conference during Apollo 7's mission raised her hand and said, "I don't understand about the balls." All of the male reporters laughed until they cried.

For Apollo 8, the word went out that "ball" would return to a gentleman's "zero." The order could not have come from the public affairs office—no one there would have dared—and the suspicion was that it had come from Borman himself, a very different sort of man from Wally. Kids would be watching the coverage, after all, and whenever someone even remotely connected to the mission snickered about language, they were not focusing on the work at hand. Still, habit was habit and aviators were aviators, and over the course of the mission, a few balls were still slipping through.

The talk about urine posed a more difficult problem. It went deeper than mere language; besides, there was no way to avoid talking about it. Newton's laws of motion—particularly the ones about objects in motion tending to remain in motion and all actions having an equal and opposite reaction—may have been all that was needed to get a spacecraft moving toward the moon and then keep it going even after its third-stage engine had completed its work. But maintaining a true course took a little more effort.

The giant main engine—the SPS—provided a booming 20,500 pounds of thrust, and each of the sixteen little reaction control thrusters produced 100 pounds. But physics makes no distinction between forces you want and forces you don't, and a fine mist of urine or other wastewater venting from the side of the spacecraft could provide a tiny thrust of its own.

It wasn't an issue anyone had worried about during the carousel ride that was Earth orbit, but on what was supposed to be a more or less as-the-crow-flies route to the moon, even a tiny nudge off course at the beginning of the trajectory could mean a huge error at the end. Already, less than forty-eight hours into the mission, the guidance officers were noticing a slight drift in Apollo 8's path. As a consequence, they were regularly discussing with the astronauts how and when to schedule their urine dumps.

In some ways, Chris Kraft was glad of the development. Potential midcourse corrections had been built into the mission's flight plan; as needed, brief burns of the SPS and long burns of the thrusters would make sure the crew traveled nearly a quarter of a million miles and yet hit the precise point that would allow them to enter lunar orbit a scant 69 miles above the surface. Thruster firings were usually preferable, both because they were less complex than firing the main engine and because any trajectory problem that could be corrected with thrusters alone was, by definition, a relatively small problem.

But Kraft preferred—indeed, very much wanted—an SPS burn, and the drift in trajectory provided the perfect opportunity to have one. In Kraft's view, it would be recklessness of the first order if the spacecraft arrived at the moon without ever having tested the main engine, which would be needed to get the crew into and out of lunar orbit. If there was an issue with the engine, it would be better to know that now and have a couple of days to deal with it than for the astronauts to learn about it only when they tried to fire the SPS on the far side of the moon, a point at which they would be completely out of radio contact with Houston. After listening in on the trajectory chatter from his observer's console at the back of the firing room, Kraft strode up to what was known as the trench—the front row of consoles where the flight dynamics and guidance officers sat—to make his preferences known.

"We need that SPS engine to work, and I very much want to see it burn before we go behind the moon," he announced.

Some of the men on the consoles shook their heads. Kraft could see the plot board and read the telemetry as well as they could. The spacecraft was off course only slightly; it was well within the limits that called for merely burning the thrusters and leaving the big gun at the back of the spacecraft dry. A poorly executed burn of the engine would only make things worse.

"If this thing burns and it's out of attitude," one controller mustered the courage to say, "we could be off trajectory by quite a bit."

"I don't care," Kraft answered. "Fire that thing and I'll get it back on trajectory for you. But I want that engine fired before we get there."

Then he turned and left the trench, and the engineers set about doing the only thing they could do—which was whatever Kraft told them. As instructed, they ran the numbers for the maneuver and found that it would be an almost absurdly minor exercise. The necessary course correction would require a pulse of the SPS lasting slightly longer than two seconds, a job the thrusters could have handled easily.

Still, Houston called the maneuver up to the ship, and the astronauts prepared to execute it. Anders, sitting in his right-hand seat, read off the steps for a burn from the flight plan, and Lovell, in the equipment bay, punched them into the computer—knowing, from the months of simulations, what Anders would read off before he actually did it, but abiding by the call-and-response rules that governed any burn of the engine. Borman, in the left-hand seat, counted down. Then the engine lit and the ship bumped forward; 2.4 seconds later, the exercise was over.

"Like a big spring," Borman said, surprised at the sudden lurch, but otherwise shrugging it off.

Kraft reacted very differently. The second the engine was lit, he saw an anomaly in the data. It was exactly the sort of problem he'd feared, exactly the reason he'd wanted a burn. There, in the stream of telemetry monitoring the fuel lines, he saw a bump in the flow of the hypergolic chemicals as they streamed toward the combustion chamber. It didn't look like much unless you knew exactly how the data was supposed to look, and Kraft did know.

He summoned George Jeffs, North American Aviation's Apollo program director, who would be in the control room during the mission, and together they walked down to the trench. The men on the console had seen the anomaly, too. The problem was minor enough that they could have overlooked it, but the entire point of their jobs was to see everything and overlook nothing.

Kraft, Jeffs, and the engineers on the console discussed the likely cause and implications of the troubling signal, including the possibility that it was merely what was known as "instrumentation," an erroneous readout, rather than any actual problem with the spacecraft. But that was probably not the answer, and in short order Jeffs figured out the real problem. Doing away with the complicated pumps that would normally feed fuel into the combustion chamber and replacing them with compressed helium that pushed the fuel along in a simpler way was one of the many innovations that made the SPS so reliable. But Jeffs realized that during a ground test of the system, not all of the helium had been properly purged from the lines. That had caused a bubbling in the fuel when the engine had undergone its brief burn, and that bubbling, in turn, had shown up in the telemetry.

The helium was now gone, vented out of the spacecraft with the exhaust when the engine lit, and therefore the fuel lines should now be entirely clear. Kraft nodded in satisfaction, cast a stern look at his trench controllers, and returned to his observer's console. Watching every single thing going on in the room was exactly what he would continue doing until the three astronauts were safely home.

☆ ☆ ☆

Nobody had expected the Apollo 8 command module to be anything like a pleasant place throughout the mission, given that three grown men would be confined in it for six full days. But no one had expected it to get quite so ripe quite so fast. Borman's space sickness had gotten the journey off to a disagreeable start, and things had pretty much gone downhill since then.

Urine dumps would still be made as needed, but with the trajectory to worry about, the exact meaning of "as needed" would change. No longer would liquid be vented outside as soon as it left the astronaut's body; instead, it would be stored in sealed plastic bags. The bags, however, were never sealed quite well enough to prevent the singular smell, if not the urine itself, from leaking out. And for every bag that was filled and stored, the odds that one of them would be bumped and ruptured increased.

The stuffiness inside the ship was another problem. The temperature on any one part of the exterior skin of the spacecraft varied from about 200 degrees below zero to 200 above, depending on whether that section was facing the glare of the sun or the deep freeze of shadowed space. To keep things balanced, the spacecraft would spend most of the flight in what was known as passive thermal control—or PTC—mode. This was, quite simply, a slow, one-revolution-per-minute rotisserie roll that would be initiated with a single burn of the thrusters and would continue indefinitely until there was a counterthrust.

But there was no PTC for the inside of the ship, and throughout the voyage the cabin temperature hovered at about 80 degrees, thanks mostly to the sunshine periodically streaming in through the windows. The spacecraft had heaters, but beyond a fan to circulate the interior atmosphere, there was nothing to cool the air.

In addition to the excessive warmth, the astronauts' personal equipment also had problems. The booties they'd been given to wear had begun to fray almost immediately after they'd put them on, and Borman and his crew worried that a loose thread could snag on something or throw a switch inadvertently. So they shucked the booties and got by with their socks. Then there were the headsets. When the astronauts had been sealed in their pressure suits for liftoff, they'd worn black-and-white fitted hoods under their helmets, with the earpieces and microphone built right in. The hoods were a precaution against a possible hazard: especially during launch, an ordinary headset might shake loose, and an astronaut could hardly open the faceplate of his pressurized suit to put the headset back in place. The hoods solved that problem and were immediately and affectionately nicknamed Snoopy hats, after the popular beagle in the *Peanuts* comic strip.

During much of the flight, headsets similar to the ones the mission controllers wore would be fine—or they would have been, if they hadn't proven impossibly glitchy, with communications regularly fuzzing in and out and static in the background often making the voice that did manage to come through the earpiece unintelligible. Before long, the astronauts ditched the headsets, retrieved the soft Snoopy hats from the lower equipment bay, and wore them throughout the mission, even during TV broadcasts. That might have cost them a measure of dignity, but it gained them a lot of function.

Getting adequate rest was often an issue for astronauts, although fortunately both Borman and Lovell found that sleep came relatively easily, or more easily than it had in the Gemini spacecraft, where there'd been no place to stretch out and where one astronaut had been required to stay awake so that the systems would be monitored and voice communications with the ground could be maintained at all times. In the Apollo spacecraft, many of the operations were automated, so Houston allowed the astronauts to sack out simultaneously if they wanted to, and the lower equipment bay included sleeping bags that could be rigged like hammocks.

But Borman did not care for the idea of simultaneous sleep shifts. It felt like negligence, like falling asleep at the wheel of a speeding car, even if there was nothing remotely nearby to hit and no accelerator or brake to mind. So on the second night in space, the commander forced himself to stay awake while Lovell and Anders slept. At about the forty-hour point in the mission—it was 11 p.m. in Houston—Jerry Carr, who was manning the capcom console, tried to keep Borman company.

Carr passed on more details about the release of the *Pueblo* crew, telling the astronaut about the thirty minutes it had taken the men to cross the aptly named Bridge of No Return, which ran from the North to the South, one man every twenty seconds or so.

"They started to cross at about eleven-thirty a.m. and were over by noon," Carr reported, keeping his voice low so as not to disturb Lovell and Anders. "On ball scores, did you get the word on Baltimore and Minnesota?"

"Not the final one," Borman said, matching Carr's quiet tone.

"Final score was Colts twenty-four, Vikings fourteen. That gives them the Western Conference, so it looks like for the NFL title it's gonna be the Browns versus the Colts on the twenty-ninth." If the Colts won that game, Borman knew, they would go on to the third annual Super Bowl game against the American Football League champions, possibly the much less experienced New York Jets.

"About the only other thing is the weather," Carr said. "It is cold, good visibility, and it's beginning to feel like winter again."

"Good time for Christmas," Borman noted idly. "Good weather for Christmas."

"Frank, we had a little eggnog over at Charlie Duke's tonight," Carr reported. Duke was a young astronaut who had never flown, but he was considered to have a fair shot at a lunar landing very late in the program. "Val Anders dropped by. She's looking fine. Tell Bill she's doing real fine."

"Fine," Borman echoed.

After a brief pause, Borman thought of the late hour in Houston and asked, "How do you like shift work, Jerry?"

"It's great, Frank."

Borman fell silent and looked out his window. The home planet, so much farther away than it had been during the first broadcast, hung before him.

"Boy, Jerry, that Earth is sure looking small."

"Roger," Carr answered. "I guess it'll get smaller, too."

Borman smiled. "Yes," he answered softly, "we're getting along pretty good."

☆ ☆ ☆

Milt Windler didn't tell a lot of people about the little flannel flags he was making for every man in Mission Control. Actually, he wasn't making them himself—he was a flight director, after all, and flag making wasn't precisely in his range of skills. His mother was doing the work. She was visiting Houston for the holidays, was handy with a sewing kit, and liked the idea of being able to contribute in even a tiny way to the historic mission in which her son had so big a role. The flags would be red and blue and would have a prominent number 1 in white in the center. Milt insisted that everyone in Mission Control had to get one, but that was a lot of flags, so some of the wives of the other controllers had agreed to pitch in and help.

The question of how the flags would get used—exactly when they would get waved—was still to be settled, however. Mission Control had a long-standing end-of-mission tradition: cigars were handed out at the beginning of the splashdown shift, though there were some very strict rules about when they would be lit. The moment the spacecraft hit the water was too early; recovery could still go south, as it had when Gus Grissom's Mercury capsule flooded with water and sank and Grissom himself, flailing in the Atlantic, almost followed it down.

Even the helicopter ride to the carrier seemed premature; after all, helicopters occasionally crash. But once the crew was on the ship, out of the helicopter, with their feet on the deck, it was safe for Houston to celebrate. If you couldn't trust a naval vessel to bring your boys home, what exactly could you trust? So that was the moment when the matches in Mission Control flared and the cigars got lit.

Windler didn't think they needed to wait that long for the flags. For him, the critical point in the mission—the point that shouldn't be passed without being marked in some way—would be when the astronauts had completed their ten orbits of the moon, fired the SPS for their trans-Earth injection, or TEI, and begun speeding home. That would be the moment the Russians were effectively beaten, and like many of NASA's men, Windler very much liked the idea of the Russians being beaten.

When Windler had first thought up the idea of the flags, he'd considered buying a sufficient number of the simple, handheld versions of the Stars and Stripes, which he could have easily picked up in any five-and-dime store. But on reflection, they seemed too generic, and that was when he came up with his plan for the custom flags. Soviet government officials would surely see the pictures from inside Mission Control—NASA would release them as widely as possible—and the bright white number 1 on the flags would be unmistakable, especially to the leaders of the country that would be so decisively reduced to number 2.

So on the evening that Jerry Carr was staying up late to chat with Frank Borman, Windler, who was off shift and rightly should have been getting some sleep, was at home, helping his wife and mother cut flannel and thread needles. It was a silly way to pass the evening for a man who spent his day managing a roomful of men who were sending a spaceship to the moon, but for Windler it seemed like the right way, too.

☆ ☆ ☆

About eight hours into Apollo 8's third full day in space, a line would be crossed. The crew wouldn't see it happen, and they wouldn't feel it happen—they'd still be soaring silently through space with more than a half a day to go before they reached the moon—but it would happen all the same. At precisely fifty-five hours, thirty-nine minutes, and fifty-five seconds of mission-elapsed time, the astronauts would cease being people of the Earth and instead became people of the moon.

Hour by hour, the spacecraft was losing its hard battle against the Earth's gravity. The velocity record set by the astronauts when they peeled off for the moon at 24,200 miles per hour had undeniably been thrilling, but the speedometer had begun drifting down almost immediately. By the 190,000-mile point in the outward journey—or 3:30 p.m. eastern time on Monday, December 23, about the time the second live television show was scheduled to be broadcast—Apollo 8 would be creeping along at only 3,800 miles per hour, or less than 16 percent of its original speed.

But then, an hour or so later, as the spacecraft passed the 200,000-mile mark, the power balance between the parent planet and its little moon would shift. At that point, the increasing pull of lunar gravity would at last overcome the greater but fading pull of the Earth's gravity, meaning that the uphill march would become a downhill plunge and the spacecraft's speed would once again begin increasing. After that moment, it was physically inevitable that Apollo 8 would at the very least whip around the far side of the moon. Perhaps the crew would succeed in entering orbit; perhaps their engine would fail them and they would be tossed back home by lunar gravity; perhaps the engine would overburn and they would crash into the moon. But whatever happened, Borman and his crew would become the first humans in the history of the species to see the lunar far side.

By the morning of the mission's third day, the ship's growing distance from the Earth was showing itself in other ways. For one thing, the sun began playing a new role: as the spacecraft's position relative to the center of the solar system continually changed, more and more sunlight streamed through Apollo 8's windows. The flood of light made a mess of Lovell's navigational optics.

Proper plot mapping meant roughly sighting stars with the system's small navigational telescope and then precisely triangulating on them with a sextant. But with the most prominent star of all—the sun—overwhelming the telescope, Lovell was left to work with the sextant alone. That was possible but painstaking, a little like using a yardstick to measure miles.

Borman reported the problem to Collins, who was back as the capcom, and Collins asked what the particular sextant angle was at the moment. Lovell, who was listening in on the conversation from the navigation station, called the answer up to Borman.

"Substellar 33," he reported.

"Substellar 33," Borman repeated for Collins.

"Roger, understood," Collins said.

Collins didn't say so, but Borman's answer was troubling. Substellar 33 should have been a clear point for good sightings. The fact that it wasn't meant that the crew would find it increasingly difficult to navigate as the time for lunar orbit insertion approached. It was one more detail the mission planners would have to address before future crews came back to land.

Collins scanned his own navigational readouts, trying to get a better sense of how the increasing amount of sunlight would affect Lovell's eyeball sightings. As he did, he noticed another intriguing bit of data. Though it was entirely predictable, it still made him smile.

"Just a matter of interest," he said. "It is taking your voice about one-point-six seconds to get down to us." The light-time distance to Earth was showing itself more and more.

"I'm a little hoarse, that's why," Borman joked.

Collins did have one more issue to discuss with the crew before he could let them get back to the business of minding their spacecraft and preparing for the television broadcast. He didn't particularly look forward to broaching the topic, but Cliff Charlesworth, who was now at the flight director's console, had ordered him to bring it up.

"When you have a moment," Collins said, "we'd like to hear a detailed crew status report from you."

"Like what?" Borman asked. His tone was not encouraging.

"Well, like, we would like to know, in the last twenty-four hours, has anyone had any symptoms similar to Frank's?"

Referring to Borman in the third person was shrewd: Collins was making it clear that he was asking about the health of the crew as a whole, not just that of the previously infirm commander, who would not take well to being treated as somehow less fit to serve than Anders and Lovell. The flight surgeon wanted Collins to remind the astronauts that they had medications on board—Marezine for nausea and Seconal for sleep—and that they should take them as needed.

"We would also," Collins went on haltingly, "like to know . . . you know, we told you the other day to take Marezine as you like. We would like to know if anyone has taken any drugs."

"Okay, nobody has taken any other drugs," Borman said, a harder edge audible in his voice. "Nobody took any Marezine; nobody is sick. Everybody had breakfast this morning and had most of a meal. What else do you want?"

This was not getting easier. "We would like to tell you to drink plenty of water," Collins said. "We think that your water intake may be down."

"We all feel fine," Borman answered, hoping to close down the discussion. Then Anders floated over to him and muttered something the ground could not hear. Borman rolled his eyes.

"I stand corrected," he said to Houston. "William had one Marezine. He didn't tell me about it; he snuck it."

"That's Bill Anders," Collins said, knowing full well who it was but saying it aloud all the same because he couldn't know if the networks were currently streaming the air-to-ground conversation to the world. Though whether the world was listening in or not, the medical interview was now over.

☆ ☆ ☆

Several hours later, it was time for Apollo 8's second broadcast. Monday at 3:30 p.m. would normally be a poor slot for any TV show, but it was December 23, schools were closed, and plenty of businesses were already shuttering for the holidays.

The first broadcast had been aired during a far more favorable Sunday afternoon hour, but not only had it gone up against an NFL game, it had faced stiff competition in the El Lago neighborhood in Houston, where so many of the astronauts had their homes. Every year during Christmas week, a fire truck would ride through the community with

Santa Claus in the back. When this year's appearance had been scheduled, no one could have known that it would coincide with the moment three of the men from the neighborhood would be broadcasting greetings from halfway between the Earth and the moon.

"One little homey item," Carr had told Borman the night before the second broadcast. "In the El Lago area, you were upstaged by Santa Claus. So most of the little critters were all outside."

Today the critters and their parents would be in front of the screen, and when the broadcast began, it was immediately clear that the astronauts would be giving their audience a much better show. Progress in the mission meant progress in the quality of the pictures the crew was able to send home. The planet outside the spacecraft's window was still grainy and without color, but a proper polarizing filter sharpened the image considerably, as did simply holding the camera still. This time, the blurred picture of the Earth seemed almost to sizzle into focus.

Before the broadcast began, Borman fired his thrusters to stop the rotisserie roll so that Anders, who would be handling the camera again, could fix it squarely on the Earth. Then Borman looked out at the view and was struck by what he saw.

"We are looking at the Earth right now, and there is a spectacular long, thin band of clouds," he said. "It's absolutely spectacular, going all the way—or almost halfway—around the Earth."

"Roger, we would like you to repeat that during the TV narrative," Collins said, "and we would like you, if possible, to go into as much of a detailed description as you poets can possibly muster."

Poetry would not be quite what the audience would get today. But if the people viewing the broadcast looked hard and squinted properly, the slightly better picture made it possible to actually see the clouds and landforms of Earth that the astronauts were describing.

When the show began, Lovell played narrator. "What you're seeing," he said, "is the Western Hemisphere; at the top is the North Pole."

After allowing that information to register, he continued: "Just lower to the center is South America, all the way down to Cape Horn."

After another pause, he said, "I can see Baja California and the southwestern part of the United States."

Whether the audience could actually make out the shapes described by Lovell wasn't the point. It was the fact that they had the chance to try, to see their home planet from a remove of 200,000 miles and make of it what they could, just as the astronauts were doing.

Exactly as Houston wanted him to, Lovell went on to talk about the Earth's colors, describing where the brown of desert gave way to the blue of water and the white of clouds. Then, at least a little transported by it all, he gave the ground a bit of the poetry it had requested.

"What I keep imagining is if I am some lonely traveler from another planet, what I would think about the Earth from this altitude," he said, "whether I would think it would be inhabited or not."

"Don't see anybody waving, is that what you're saying?" joked Collins. After having asked for the lyricism, he now seemed to be laughing it off.

But Lovell would not be denied a whiff of wonder. "I was just kind of curious if I would land on the blue part or the brown part of the Earth," he mused.

"You better hope we land on the blue part," Anders said.

"So do we, babe," Collins answered.

Lovell smiled and went back to the more straight-up work of describing without reflecting. Privately, however, he decided that he was at that moment about as happy as he'd ever been. And more than ever, the idea of finishing this mission and flying the whole thing all over again—only this time with a LEM, so he could land on the world they would begin orbiting tomorrow—seemed very, very appealing.

The broadcast finished about seventeen minutes after it began. The end came without any farewell from the crew: just short of the time the show was scheduled to be completed, the TV signal was suddenly lost. But there would be plenty more video to come, including two more broadcasts during the twenty or so hours the crew would spend in lunar orbit. For now, though, the astronauts stowed the camera, restored the rotisserie roll, and went back to the routine business of flying their ship without hundreds of millions of people watching them do it.

Before long, Collins called up to them once again. "By the way," he said, "welcome to the moon's sphere."

"The moon's fair?" Borman asked, not quite making him out through the static.

"The moon's sphere," Collins corrected. "You're in the influence."

Silently, invisibly, the gravity of the Earth—in the grip of which the human species had always lived—had handed Apollo 8 off to the gravity of the moon. The crew's long lunar plunge had begun.

THIRTEEN

December 24, 1968

IF THE HOUSTON OILERS HAD BEEN A BETTER FOOTBALL TEAM, Susan Borman would have had less to think about as her husband was preparing to sail his spacecraft around the far side of the moon. The Oilers, however, were perfectly ordinary, which left her sons with nothing to do during Christmas week.

Both Fred and Ed Borman worked for the Oilers in their off-hours, partly because of their famous father and partly because they had inherited his strict sense of professional discipline, which was evident to anyone who met them. The Borman boys, seventeen and fifteen years old, respectively, had begun their work for the team as ball boys, but they'd performed so well that they had quickly been promoted to co–equipment managers, more or less the same football job their father had held at West Point.

During the 1968 season, the Oilers won seven games and lost seven, which put them four games behind the first-place New York Jets. Had the Oilers been a little bit better, they would have had a playoff game scheduled for December 22—the day after Apollo 8 lifted off—keeping the Borman boys out from underfoot and, more important, keeping their minds occupied, in case they were inclined to worry about whether their father would make it home from his trip to the moon. But the Oilers did not make the playoffs that year, which meant Fred and Ed were home the entire week, giving Susan one more thing to attend to as her house remained filled with people and her husband sped farther and farther away. So she did the only thing she could think to do for her boys, which was to make their week as unremarkable as possible.

Since Susan would often get up early on school days and cook the boys pork chops and eggs for breakfast, she did that during this holiday week, too, the smell sometimes waking several of the astronauts and their wives, who had fallen asleep in her living room while watching the coverage of the mission overnight. For the guests it was the cue to go home, shower, make their own breakfasts, and come back later in the morning.

For Fred and Ed, breakfast gave them a chance to enjoy some brief private time and eyeball their mother to gauge how she was holding up. The answer, they could readily see, was not terribly well. As always, she took good care of her guests and smiled gamely for the press, so her picture in the papers would look fine if you didn't know her very well. But Fred and Ed could read her face with a fluency almost no one else had, and everything from the tight set of her mouth to the spooked look in her eyes spoke to them of a kind of knotted terror.

Fred, the older boy, took it upon himself to ensure that he and Ed behaved this week. *Let's not act like knuckleheads* was the way he liked to put it. The bigger problem, of course, was the simple matter of boredom, with every second of every day filled with nothing but moon talk. Having grown up with the space program, they knew no other life, and it took a lot to get them excited. They had watched the Apollo 8 launch along with the adults, although when the time for the burn out of Earth orbit arrived, they drifted from the room. They could hear their father's voice coming through the squawk box and the TV, and during the two broadcasts back to Earth, they wandered into the living room, got a peek of him on the screen—he was growing a beard, but with his sandy hair it was less noticeable than Jim Lovell's and Bill Anders's dark scruff—and then they wandered back out.

When they could, they left the house and went over to their friends' homes, where there were no crowds and no newsmen and the TV might not be on at all. "Going out," Fred would shout to Susan as they were getting ready to leave, taking care not to go into the living room, where they would have to say good-bye to a dozen people they hadn't even said hello to earlier that morning.

"Not the front door!" Susan would call back, an unnecessary reminder that the front yard was jammed with reporters and photographers.

The backyard was safer, surrounded by a high fence with a door that had a sturdy latch, which meant that unlike Valerie Anders, the boys had no reason to fear that a cameraman would be waiting in the yard to ambush them. And whenever they suspected that someone was lurking on the other side of the fence door, they would avoid it entirely, vault the farthest stretch of fence, and take the back streets to their friends' homes.

When they arrived, an adult would inevitably ask about the mission, to which they typically had little to say.

"Not really thinking about it," they would answer.

And if anyone raised the topic of the danger their father faced— which too many people seemed inclined to do—Ed would remind them who they were talking about.

"Dad's in the Air Force," he'd say, "a lieutenant colonel. There's a Cold War going on."

Ed Borman was still only in high school, but he sounded so much like his father when he said such things that the adults who heard him could only shake their heads.

Their father, off fighting that war, would have approved.

☆ ☆ ☆

Anyone who hoped to mark the moment when Apollo 8 entered lunar orbit would likely be disappointed. The spacecraft was set to arc behind the moon at 3:50 a.m. Houston time, in the first few hours of the morning of Christmas Eve, breaking the communications link between the spacecraft and Mission Control that had been maintained since the astronauts had climbed into their ship three days earlier. On the screens in

Houston, it would be as if the Apollo 8 spacecraft and its crew had simply vanished into space. The data stream would be reduced to nothing—all balls—and only a translunar hiss would fill the headsets.

The communications blackout would last about thirty-five minutes. Ten minutes into it, the astronauts would fire the SPS engine for the lunar orbit insertion, or LOI, burn, and that either would or would not make them satellites of the moon. Until they emerged back into the storm of radio waves streaming from their home planet, nobody on that planet would know if the three men were dead or alive.

The engine burn that would settle the matter one way or the other would have to be executed very precisely. As the spacecraft made its final approach before loss of signal, or LOS, its 3,800-mile-per-hour speed would increase to 5,800 miles per hour, thanks to the pull of the moon's gravity. The ship would be flying backward, with the engine bell facing forward so that the burn—due to be a long one, lasting four minutes and two seconds—could act as a brake, bringing the speed back down to 3,700 miles per hour. That was slow enough for lunar gravity to grab the ship and settle it into orbit.

Those were the nice round numbers that NASA gave the press, but the astronauts and the mission controllers preferred to work in units that were more precise, reaching far to the right of the decimal point. Fudge your figures even a little and everything might come apart, which was the nature of things when you were traveling more than 233,000 miles to enter an orbit as low as 69 statute miles above the lunar surface—which sounded even more more perilous when expressed as the 60 *nautical* mile measure NASA preferred to use. Whether nautical or statute, it meant a margin of error of just .0296 percent, the equivalent of standing at one end of a football field and shooting at an apple in the opposing end zone—but aiming to skin it, not hit it.

Gerry Griffin, who worked the guidance and navigation console in the trench, liked to joke about how precise the numbers had to be. "They must have a lot of faith in us," Griffin would say, "trusting that our numbers will put them into orbit sixty miles above the surface instead of sixty miles into it." But even the teller found the joke only darkly funny.

For the astronauts themselves, the final approach to the moon would be an exercise in flying more or less blind. They hadn't glimpsed the moon since their first day in space, when they had been traveling nose forward. Even then, the prow of the ship had often been inclined either above or below the lunar plane—depending on the navigation readings that needed to be taken—meaning they saw only stars through their windows. And once they had turned around to face the Earth for the second TV show, they had simply remained that way in preparation for the LOI burn.

"As a matter of interest," Lovell called down at 2:50 a.m. Houston time, an hour before loss of radio contact, "we have as yet to see the moon."

"Roger," said Jerry Carr, who was now at the capcom console. "What else *are* you seeing?"

"Nothing," Anders answered, looking at the windows, most of which were stubbornly fogged. "It's like being on the inside of a submarine."

"Roger," Carr responded.

Without needing to ask, Carr knew that Anders was talking about the windows, and he knew that none of the astronauts were happy with them. The engineers hadn't anticipated that the fogging would be as bad as it was, but after examining the problem over the past couple of days, they had concluded that it would probably correct itself. When the spacecraft entered lunar orbit, the reflected sunlight from the surface of the moon would raise the temperature on the skin of the ship by as much as 50 degrees, or at least that's what the flight planners promised would happen. If they were right, the higher temperature would change the density of the gas trapped between the window frames and melt any frost on the panes as well, thus improving the view.

As the time to loss of signal and the lunar orbit insertion burn slowly ticked away, both the crew in the ship and the men on the ground attended to minor housekeeping chores, confirming communications settings, cockpit recorder status, trajectory details, and the like. But it was really little more than busywork, a way to remain occupied during the final approach to the moon.

Finally, at sixty-eight hours and four minutes of ground-elapsed time, the capcom gave the crew the official approval for the mission's riskiest moment, the maneuver on which the entire voyage depended.

"Apollo 8, Houston," Carr said, "at sixty-eight-oh-four you are go for LOI."

"Okay, Apollo 8 is go," Borman answered.

Carr sat uneasily in the silence that followed. As Collins had found, the cold jargon of the capcom was simply not up to the magnitude of what he had just given the crew the clearance to do.

"Apollo 8, Houston," he finally added. "You are riding the best bird we can find."

"Say again," Borman requested.

"You are riding the best bird we can find. Over."

"Thank you," Borman replied. Then, five seconds later, he added a small bit of sentiment of his own: "Roger, it's a good one."

As if on cue, Mission Control's big map on the wide screen at the front of the room changed: after showing the translunar route for the better part of the past three days, it switched to a lunar map. The path the spacecraft would follow if the burn went well was traced across its surface.

"Apollo 8, Houston, we have got our lunar map out and ready to go," Carr told Borman.

"Roger," Borman said.

After that, there was little chatter until about forty-five minutes later, when Carr once more hailed the ship.

"Apollo 8, Houston," he said. "Five minutes till LOS; systems go, over."

"Thank you," Borman answered. The static crackled, and then Carr spoke up again.

"Frank," he said, "the custard is in the oven at 350."

"No comprendo," Borman said.

And then, an instant later, he did understand. It was a message from Susan, who, if he knew her at all, would be keeping the house together and keeping the guests looked after but would, as the loss of signal and engine burn approached, be listening to the squawk box alone somewhere, in the kitchen perhaps, maybe in the bedroom, with no one beside her to see her relief if things went well or her devastation if they didn't. As Susan knew, public displays of powerful emotions were simply not how things were done.

Still, she could send Frank a message. So she had spoken to Carr when he was off shift and asked him to radio up the coded declaration of family solidarity, the incantation they had shared since the earliest days of their marriage, when she would promise him that she would let him handle the risky business of flying if he would let her handle the household and the children.

Now Borman smiled—a small, private smile. Perhaps Anders caught it, perhaps he didn't. Either way, he answered for his commander, preventing Carr from repeating the message and spoiling whatever its secret sentiment was.

"Roger," Anders said.

Shortly afterward, Carr announced the one-minute mark to LOS. Then, with only seconds remaining, the man at the capcom microphone spoke the words everyone else in the control room was thinking. "Safe journey, guys," he said.

"Thanks a lot, troops," Anders answered.

"We'll see you on the other side," Carr called.

A moment later, at exactly the second NASA had calculated, the signal cut out.

The men in the spacecraft looked at one another partly in wonder, partly in admiration.

"That was great, wasn't it?" Borman asked. With a laugh, he added, "I wonder if they just turned it off."

Anders grinned. "Chris probably said, 'No matter what happens, turn it off,'" he added.

Chris Kraft had said no such thing, of course, and the Apollo 8 crewmen knew it. As the long minutes of radio silence began, the three astronauts were disconnected from the rest of humanity in a way that no one ever had been before.

☆ ☆ ☆

Gene Kranz stood at the back of Mission Control and listened for the rasp of Zippo lighters, and they began almost as if on cue. Zippos were forever lighting cigarettes in the big auditorium, where most of the controllers smoked and many of them were either military men or worked closely with military men. The Zippo was a soldier's lighter—heavy, solid, all but indestructible, with a lid that opened and closed with a satisfying snap, releasing in that brief interval the signature whiff of the lighter fluid saturating its wick.

Most of the time, you couldn't hear the Zippo's familiar rasp in the room, but in quiet moments such as this one, you could. When a spacecraft orbiting the Earth passed between the footprints of the various tracking stations circling the globe, ship-to-shore contact would be briefly lost. Those silences were predictable and familiar, and they carried no special portent. But no ship had ever gone into the blackout that was caused by flying behind the 2,159-mile-wide bulk of the moon, and this silence felt different.

Chris Kraft, sitting at his observer's console near where Kranz was standing, took the opportunity to pour himself a cup of tea. He walked over to the table where the hot plate and coffeepots were kept and stood next to Bob Gilruth, not only the first director of the Manned Spacecraft Center in Houston but one of the original members of the Space Task Group, the committee established by President Eisenhower in 1958 that led to the creation of NASA.

"Ten years and a month," Gilruth said, referring to exactly how much time had passed since he'd written the first task force memo. "Ten years and a month. There were thirty-six of us on the list, including you, Chris."

"And now we've got three men behind the moon," Kraft answered. "If we weren't sitting here, I don't know that I'd believe it."

Both men gazed at the mission clock at the front of the room. The length of the blackout was not absolutely fixed. If all went well, it would last about thirty-five minutes—a bit longer if the engine fired successfully and the ship slowed to 3,700 miles per hour and settled into orbit, a bit shorter if the engine failed to fire and the ship continued to speed along at 5,800 miles per hour. If something worse happened, the radio silence would last forever.

☆ ☆ ☆

Inside the spacecraft, the astronauts were too busy to spend a lot of time thinking about all of the possibilities facing them. The burn was less than ten minutes away, and there was a lot to configure before the ship would be ready.

They spoke with one another using the technical patter familiar to pilots and mariners throughout history. One crew member would call out settings, and another would respond with confirming echoes.

"Deadband minimum," Anders announced, checking a stabilization system.

"Deadband minimum," Borman repeated.

"Rates low."

"Low."

"Limit cycle, on."

"Limit cycle, on."

"Have you got the GDC aligned?" Anders asked, referring to a part of the gyro display system.

"Yes," Borman answered. Five seconds later, he repeated himself—the proper way, this time: "GDC aligned."

The final configuring was supposed to take up four or five of the ten minutes that remained, but the crew had drilled the procedure so many times on Earth that they completed it in just two of those available minutes.

"Okay, eight minutes," Borman said, glancing at the countdown clock.

Lovell looked around himself. His crewmates did not seem tense, but they didn't seem especially relaxed, either. "Well, the main thing is to be cool," he said.

He gave the radar indicator a look. With the spacecraft windows facing up and the ship pointing backward, the astronauts still couldn't see the moon. But the instruments knew that it was now directly beneath them.

"Well, I'll tell you, gentlemen," Lovell said, "that moon is pretty close."

"Seven minutes," Borman said.

A minute later, he called off six minutes, then read off a few switch positions for Anders to confirm. A minute after that, Lovell called off five minutes.

Borman squinted out the window and frowned. There was still only blackness. "On that horizon, boy, I can't see squat out there," he said.

"You want us to turn off your lights and check it?" Anders asked, reaching for the interior lighting switches.

Before he could touch them, however, Lovell called out: "Hey, I got the moon!"

"You do?" Anders asked.

"Right below us!" he said, looking through his window.

And indeed, there it was, just visible through Lovell's backward-facing window. The ship had now traveled far enough past the leading edge of the moon that some of the ancient gray surface stretched beyond the Apollo's nose. The expanse was huge—a meteor-blasted beach that spread out to the right and left until it spilled over the horizons.

Jim Lovell, the first human being in history to see the far side of the moon, stared at the ruined landscape, transfixed and momentarily speechless.

"Is it below us?" Anders asked excitedly, pressing close to his window.

"Yes, and it's—" Lovell began.

"Oh my God!" Anders exclaimed.

"What's wrong?" Borman said.

But nothing was wrong. Anders was now seeing the moon for himself. "Look at that!" he said. "I see two—" He waved his hands to fill in the word "craters," which, in his excitement, was eluding him. "Look at that!" he repeated.

"Yes," Lovell said.

"See it?" Anders said. "Fan . . . fantastic."

The position of the sun relative to the ship, which would change continuously as the spacecraft sped along, was similar to that of high noon on Earth, and the direct light washed out any lunar shadows, making it hard to distinguish a concave crater from a convex hummock.

"I have trouble telling the holes from the bumps," Anders said, still struggling to reclaim his geographer's jargon.

"All right, all right, c'mon," Borman scolded, minding his ship while his crewmen gaped. "You're going to be looking at that for a long time."

"Twenty hours, is that it?" Anders asked, knowing full well that ten orbits averaging a little less than two hours each would mean they'd be looking at the moon for about twenty hours. All the same, he seemed eager to reassure himself of his sublime good fortune.

Anders and Lovell settled back into their couches, and then all three men loosely fastened their seat belts. As with the earlier translunar injection burn, the thrust from the lunar orbit insertion would be comparatively small, so the belts were meant not to keep them secure when the engine fired but to prevent them from drifting up and away.

"One minute," Borman said. Then he added, "Come on, Jim, let's watch it real good."

Lovell would do exactly that, since it was his job to work the computer keypad on the instrument panel and punch in the final commands for the four-minute engine burn. He checked the altitude and velocity readouts he was being fed by the computer and then entered the commands he had practiced a thousand times in simulations.

"Standing by for engine-on enable," Anders said. "Proceed when you get it."

"Okay," Lovell answered, entering the final keystrokes.

With ten seconds to go, he consulted the computer readout one more time. It flashed "99:20" back at him. That was the computer's last-chance code, its way of confirming that the human being at the switch really did want to do what he was about to do.

When the countdown clock reached zero, Lovell hit the button marked PROCEED.

Somewhere behind the astronauts, the engine bell began emitting a silent roar of exhaust, braking the ship with 22,300 pounds of thrust. Borman, Lovell, and Anders heard none of that, but they felt it in the form of a subtle pressure at their backs.

"One second, two seconds," Borman read aloud as the first moments of the 242-second burn played out. "How's every—"

Anders, anticipating Borman's question, responded to it before he could finish. "We got them. Pressure's holding good."

"Fifteen seconds," Borman said.

"Pressure coming up nicely," Anders assured his commander.

"All right."

"Everything is great."

The engine burned on, and the ship continued to slow. Although the astronauts knew that the length of the firing had been calculated again and again by the trajectory planners, they also understood that a basic rule of space travel was that when you're going fast, you never want to slow down more than you should—at least not in the vicinity of a mass like the moon. Lose too much speed and any planned orbit would collapse, ending in a free fall.

"Jesus, four minutes?" Borman muttered as they approached the two-minute mark.

"Longest four minutes I ever spent," Lovell said, after nearly another minute had elapsed.

As the engine continued to burn, the astronauts could feel its effects growing stronger, the pressure at their backs increasing, turning into a shadow of gravity. It was less than a single g, but for men who had been weightless for three days, it felt like much more.

"It seems like about three g's," Anders said.

Lovell kept his eyes on the clock and announced when there was just one minute left, then forty-eight seconds, then twenty-eight seconds.

"Stand by," he said.

"Okay," Borman answered.

"Five, four, three, two, one," Anders read off.

And then, right on schedule, the SPS—the service propulsion system that had served, and propelled, and had now slowed the spacecraft too—went silent.

"Shutdown!" Borman announced.

"Okay!" Lovell said with a satisfied nod.

In that moment, Apollo 8 became a satellite of the moon. The spacecraft was circling it in an elliptical orbit with—as the ship's instruments confirmed—a high point of 160 nautical miles and a low point of just 60, or precisely the parameters Houston had planned. And the three astronauts now orbiting the moon were the only people on or off the Earth who knew they had succeeded.

☆ ☆ ☆

The silence in Mission Control stretched on, as the idle people at the consoles tapped their pencils, jiggled their feet, stubbed out their cigarettes and lit new ones. Many of the controllers stared at the mission clock at the front of the room; with their console screens still receiving no information at all from the ship, the clock was the only functioning data point left. So they watched it tick down, first toward the too-early moment when they would hear from the crew if the engine had failed to fire, then to the later, happier one that would indicate the spacecraft had slowed and was in lunar orbit.

When that first moment passed, a few controllers glanced at one another. But they did not dare smile yet; after all, the silence that portended good news might never end, which would mean very bad news. A few more moments went by as their headsets continued to produce only static and their screens continued to offer no new data.

Thirty-four minutes and two seconds after the spacecraft disappeared into blackout, Carr began hailing the ship.

"Apollo 8, Houston. Over," he said, knowing that it was a bit too early to expect an answer, though not too early to hope for one.

"Apollo 8, Houston. Over," he repeated thirty-three seconds later.

"Apollo 8, Houston. Over," he repeated after fifteen more seconds.

He said it yet again after eighteen more seconds, and then thirteen more seconds and then twenty-three seconds.

At last, eight seconds later—thirty-five minutes and fifty-two seconds after LOS—the lifeless crackle in the controllers' headsets was replaced by something that was very much alive.

"Go ahead, Houston, this is Apollo 8," said the unmistakable voice of Lovell. "Burn complete. Our orbit 160.9 by 60.5."

"Apollo 8, this is Houston. Good to hear your voice!" Carr answered.

In truth, the capcom barely heard a thing, since all the men around him had leapt up as one, whooping, cheering, and embracing one another in relief and jubilation. The display went way beyond what Mission Control decorum permitted. But nobody cared: concern for propriety was lost, and that was just fine. Paul Haney, the NASA commentator who was narrating the events for the early-morning broadcasts, made the obvious official.

"We got it! We've got it! Apollo 8 is now in lunar orbit," he shouted over the din. "There is a cheer in this room!"

It was much quieter in the kitchen at the Borman home, where

Susan, as Frank had suspected, had been waiting out the long lunar silence alone. But even there, the cheers from her living room were impossible not to hear.

☆ ☆ ☆

America awoke slowly to the news that three of its countrymen were in orbit around the moon. The predawn hour of the orbital burn meant that most of the people living in the nation that had sent the spacecraft on its journey were not awake when Lovell spoke the words so many had been waiting to hear. That was not the case in Houston, of course, where half the households seemed to have their lights on at 3:50 in the morning. And it certainly wasn't the case in the NASA neighborhoods like Timber Cove and El Lago, where almost nobody was in bed but the children. For most of the rest of the country, however, the happy news of Apollo 8's lunar orbit would be consumed hours after the fact, with the morning coffee.

In much of the rest of the world, which was considerably deeper into its day, the response was different. Stories about the mission had been blanketing the BBC in Great Britain, and its reporters had been covering the LOI burn live. The news that the burn had been successful arrived at the more seemly hour of 9:50 a.m. Greenwich mean time. The TVs were on in the London pubs, attracting crowds of people who pressed inside and stared up at the screens. And if they treated themselves to a pint at so early an hour, well, there was something to celebrate, wasn't there? As they watched, the country's most famous astronomer, Sir Bernard Lovell of the Jodrell Bank Observatory—no relation to the now vastly more famous astronaut—called the mission "one of the most historic developments in the history of the human race."

Television networks all over Western Europe followed the BBC's lead, interrupting their morning and afternoon broadcasts with news from the moon. Holland, for reasons known only to the Dutch, had a special hunger for the story, and even the radio stations that usually broadcast only music suspended or interrupted their programming to carry regular updates on the mission. Iran's state-controlled television network covered the story steadily and live, as did Tehran's main radio station. In Libya, a nation that had been established seventeen years ago to the day, huge rallies celebrated both the country's independence and

Apollo 8's astonishing feat. Accounts of the successful orbital maneuver were even reported faithfully in the Soviet Union. The news was not announced there until an hour and fifteen minutes after Apollo 8 sailed back around the near side of the moon, but by the standards of the state-run Soviet press, that was almost as good as live.

☆ ☆ ☆

If the crew of Apollo 8 hoped to take some time to reflect on their remarkable circumstances, they would have to do so during the quiet intervals when circling the far side of the moon. For each of their spacecraft's ten orbits, the astronauts would be in radio contact with the ground for nearly ninety minutes, and during those intervals there would be too many chores to attend to and too much chattering in their ears. Worse, a TV broadcast had been planned for the first orbit on the near side of the moon, meaning that hundreds of millions of people would be listening to and watching the goings-on in the ship. Americans who missed the broadcast due to the early hour would be able to see it on what would be a nearly continuous loop throughout the day.

During the first few minutes of blackout after the orbital burn, the astronauts had thus allowed themselves what they knew would be the brief luxury of mere sightseeing. The flight plan called for the spacecraft to roll over and fly upside down for most of its twenty hours in lunar orbit, since upward-pointing windows would be useless if you were trying to map the moon that lay below.

"Okay," Borman warned the other two when the moment for the maneuver arrived. "I've got to roll right." He took hold of the thruster handle—the same handle that the witless engineer in the Downey factory had wanted to install backward so long ago—then flicked his wrist and vented a breath of propellant through the port-facing jets. The ship immediately started to roll.

Lovell was surprised at the kick that even a small burst from the maneuvering system could pack. "Go easy on those thrusters," he warned.

After the ship spun 180 degrees, Borman executed a quick counter-thrust and the roll stopped. The windows—which, as the mission controllers had predicted, were warming and clearing—were now filled with nothing but moon. All three men peered out, and at first all three men said nothing.

Nearly 4.5 billion years earlier, a passing protoplanet had collided with the infant Earth, knocking it to the cockeyed twenty-three-degree cant that would one day give it its seasons. The collision sent up a great debris cloud that, for a short time, gave the planet a ring. Within a few eons, the ring coalesced into a moon. That new satellite would soon become gravitationally locked to the Earth, meaning that it would always keep one side facing its parent planet and the other side pointed out to space. Less than a billion years after the moon formed, the first one-celled organisms appeared on Earth. It would be another two and a half billion years before multicelled organisms, some with rudimentary light-sensitive eyespots, followed, and on dark nights those spots may have registered the photons pouring down from the brightly lit moon. Later, other organisms with better eyes—and, in one case, a mammal with a big brain and opposable thumbs and a consuming curiosity about the nighttime sky—would come along, too. And in all of that time, not a single earthly eye had ever seen the side of the moon that the six eyes belonging to the astronauts in the Apollo 8 spacecraft were seeing now.

Those men were here on a mission, and if they were given to wonder, which they surely were, they kept it mostly to themselves. But it was evident in their tone and in the long silences between their words. Those words, however, were mostly about the job they were here to do, which was to understand the science, not gawk at the spectacle.

"Boy," said Lovell softly, blinking at the bright light of the sun, which was still directly overhead and reflecting off the lunar surface. "There's no shadows in those craters at all."

He scanned back and forth across the surface, comparing the huge, dusty world he was seeing with the two-dimensional maps and photos he had been studying for so many months. Almost immediately, Lovell knew where he was. "We're passing over Brand right now," he announced, recognizing the forty-mile-wide crater that had been informally named for rookie astronaut Vance Brand, who had been on Apollo 8's support crew and had earned himself a little recognition.

Then—bigger, more brilliant—came a one-hundred-mile crater, one of the most conspicuous features left over from one of the most violent hits the lunar far side had ever sustained. "Is that Tsiolkovsky?" Lovell asked no one in particular, knowing full well it was. Studying maps of the moon and not being able to recognize the Tsiolkovsky Crater would be like studying maps of the United States and not recognizing Florida. But Lovell was asking anyway, less out of confusion than out of simple incredulity that he was here at all.

He and Borman continued to scan the ground rolling slowly below, picking out other craters, too: Scaliger, Sherrington, Pasteur, Delporte, Necho, Richardson. Once just cartographers' marks on flat pieces of paper, they were now solid formations of dirt and rock and ancient lava flows, the entire history of the moon written in its wounds. Meanwhile, Anders began bouncing around the cockpit, excitedly collecting his lenses and cameras and film magazines.

"I've got Mag D," he said, after darting down to the equipment bay and grabbing some film. "And this lens," he added, snatching up a seventy-millimeter camera and lens in one hand and displaying it to Borman and Lovell. He gathered the brackets that would be attached to the windows to hold the cameras, then reappeared carrying it all heaped in his arms.

Borman watched him warily. Three men in so cramped a space reminded him of a crowd of monkeys in a small room, and any errant move could cause a switch to be bumped or a circuit to be broken without anyone noticing.

"Hey look, just slow down. Take your time, okay?" he ordered.

Anders complied as best he could and spent the minutes until reacquisition of signal assembling his gear a bit more patiently. The twenty hours they'd be spending in lunar orbit would, he knew, go fast. And although Anders hadn't trained to be a photographer, it was the job he had been handed, and he was determined to do it fully and well.

☆ ☆ ☆

When contact with the ground was at last reestablished and the mission controllers had finished their brief, noisy celebration, the capcom and the crew spent the time they had before the television show confirming the spacecraft's orbital coordinates and the status of the SPS, which would be fired briefly during the third blackout to smooth the ship's orbit from the assymetrical ellipse that was the best the orbital insertion burn could produce to a tidy circle. Anders then unpacked the TV camera and positioned it at the window, preparing to begin the live broadcast to the portion of the world that was awake.

This time, the camera's lens would not be pointed back at the indistinct blob of a planet that, the last time the TV audience had seen it, was 180,000 miles away and growing steadily smaller. Instead, it would be taking in the surface of the moon a scant 60 nautical miles below. By then the sun's angle was growing more acute, and Anders had installed a filter that would take advantage of that fact, meaning that the craters and rills and scarps and cliffs that were now casting sharp shadows would be discernible even on television screens a quarter million miles away.

Once the broadcast from the ship began, the TV signal came first to the movie theater–like viewing screen in Mission Control. Looking at the screen from the glassed-in press booth at the back of the auditorium, the television producers could see its unexpected crispness. The spacecraft was at that point passing over the Sea of Fertility, the 522-mile-wide plain on the eastern limb of the moon formed by a vast lava bleed that had resulted from a meteor hit four billion years earlier.

Almost as one, the TV producers signaled their networks to flick the switches that would spill the feed out to their broadcast centers and, from there, to the world beyond. Jerry Carr knew that this was his signal to start the show.

"Apollo 8, Houston," he called up, affecting a jauntiness that was clearly for the benefit of the television audience rather than the astronauts. "What does the ol' moon look like from sixty miles? Over."

"Okay, Houston, the moon is essentially gray, no color," Lovell answered. "Looks like plaster of Paris or sort of a grayish deep sand. The Sea of Fertility doesn't stand out as well as it does back on Earth."

"Roger, understood," Carr answered.

Lovell then spotted two craters that had grown very familiar from his lunar maps and plots.

"And coming up now are the old friends Messier and Pickering that I looked at so much on Earth," he said.

"Roger," Carr answered.

"I can see rays coming out of Pickering," Lovell said, referring to the blast lines that had been left in the ground long ago by the impact that formed the crater. "They are quite faint from here."

"Roger," Carr repeated, but this time he sounded distracted. On the Mission Control voice loop that carried communications among the controllers themselves, he was listening to chatter about a backup water evaporator used for the coolant system that was performing poorly. That system was under Anders's command.

"Capcom, tell him to shut down the secondary evap," Milt Windler, the flight director on duty, ordered. Carr looked at the TV transmission and could see that Anders was busy snapping pictures. Turning off the evaporator was a minor task, but according to mission rules, it had to be done right away.

"Bill," Carr said, keeping his tone light, "if you can tear yourself away from that window, we'd like you to turn off the secondary evaporator."

"Roger, going off," Anders replied. He bounced away from the window, executed the command, and bounced back.

Lovell continued to narrate the features crawling past his window. After Messier and Pickering came the Pyrenees Mountains, named after the snowy, far prettier range separating Spain from France. Then, as the ship sailed over the shore of the Sea of Tranquillity, Lovell saw another landmark. The spot was significant to NASA because it might be used as an approach point for a later landing, but it mattered to Lovell for a more personal reason.

"We can see the second initial point," he said, "the Triangular Mountain."

The Triangular Mountain was the prosaic name NASA had assigned the potential approach point, but Lovell stubbornly thought of it as Mount Marilyn. He had promised his wife that he would give it that name, and now he decided to call the formation that for the rest of the mission. It might be informal, and it might cause confusion among the NASA mapmakers, but the more the name Mount Marilyn got used on the air-to-ground loop, the likelier it was to stick.

The broadcast continued for a few minutes more, taking the crew— and, by extension, the audience—over the craters Colombo and Gutenberg, and then the one-hundred-mile-wide Marsh of Sleep, with its craters Lyell, Crile, and Franz. Lovell peered ahead to the shadowed portion of the moon, which lay well forward; squinting, he tried to see if, in the absence of sunlight, the darkened hemisphere that was approaching might be discernible in the faint light reflecting off the Earth and back to the moon.

"I can't see anything in earthshine at this present time," he said.

He used the word casually, familiarly, but to the viewers it was not remotely familiar. "Earthshine"—a word that had no meaning until this very day, until humans had gone to a place where earthshine existed and mattered. It was a wonderful notion, and one the television audience might have taken a moment to savor. But suddenly, with no warning, the feed from the moon cut out.

NASA had again scheduled the moment that the broadcast would stop, but again no one at the agency had thought to prepare a proper sign-off. There would be another broadcast, during the ship's ninth orbit, and perhaps something better could be planned for then.

After the show ended, Borman was relieved to be able to get back to work. He was delighted with the job his main engine had performed, but it would be needed again for the circularizing burn and, more important, for the trans-Earth injection—or TEI—burn during the blackout after the tenth orbit. NASA had promised to keep him apprised of the telemetry readouts from the LOI burn to determine if there had been any hidden anomalies, perhaps something like the helium bump during the midcourse correction.

Borman glanced over at Lovell and Anders, who were still allowing themselves a few more moments to sightsee.

"While these guys are looking at the moon, I want to make sure we have a good SPS," he told the capcom. "How about giving me that report when you can."

"Sure will, Frank," Carr replied.

"We want a go for every rev, please, otherwise we will burn at TEI at your direction."

"Roger," Carr said, "I understand."

Anders, still at his window with his cameras, turned. He, too, understood what Borman meant. Before the beginning of every revolution, or orbit, Borman would wait for an official okay from the ground that the spacecraft was fit. If he didn't get that assurance, Borman would assume that his ship was *not* fit, and at the proper point during the trip around the far side of the moon, he would fire his engine and aim for the home planet. Borman's addition of the qualifier "at your direction" was diplomatic, since no Apollo commander would make such a mission-critical maneuver without the ground's consent unless it was truly unavoidable. But Borman also wanted to make it clear that his request for a go was not simply an idle preference, and that he would exercise the commander's prerogative whenever he felt it was warranted.

Anders, reflecting on the exchange, was glad he had set up his camera brackets so quickly. As he now realized, he couldn't count on getting the full twenty hours of shooting time he'd been promised.

☆ ☆ ☆

Apollo 8 slipped back into lunar silence one hour and forty-eight minutes after it had emerged. When the crew reappeared and began the second revolution, there would be more landmark sightings to take, particularly of some of the other landing zones NASA was considering beyond the Sea of Tranquillity. These would include the Hadley-Apennine region, the Ocean of Storms, the Fra Mauro Highlands, and the Taurus-Littrow Mountains, though for now the various sites were designated simply P1, P2, and the like. One day, however, they would be given new and far more illustrious names, something like Tranquillity Base, say, or Hadley Base.

There would also be craters to spot and to name. As with Mount Marilyn, those names would not be official unless NASA started using them and, later, the International Astronomical Union—the arbiters of all cosmic nomenclature—approved. Even for a stuffy academic body like the IAU, however, at least some of the names would be irresistible.

There would be craters named for the Apollo 1 astronauts lost in the cockpit fire—Crater Grissom, Crater White, and Crater Chaffee—as well as for Charlie Bassett, Elliot See, and Ted Freeman, all of them astronauts who'd died in airplane accidents before ever getting to space. There would be a crater for Michael Collins, who would have been on this flight but for the bone spur on his spine. There would be craters for Bob Gilruth, Chris Kraft, and Jim Webb, NASA aristocracy; there would even be one for Joe Shea, the vilified head of the Apollo program at the time of the fire. He was a banished man now, but he had served the moon cause well for years.

And, of course, there would be craters Borman, Lovell, and Anders, far in the southern lunar hemisphere, a trio of landmarks grouped closely together. As to which man's name was assigned to which crater, that would be determined by the size of the crater and the mission rank of the astronaut. So Borman got the biggest one, Lovell the next, and Anders the smallest.

Finally, some craters would be named on the fly. "Okay, we are coming up on Crater Collins," Anders told the ground during one orbit. He had turned the TV camera back on to let Mission Control follow along with what he was seeing.

"Roger," Carr said. Then the capcom noticed a smaller formation adjacent to Collins. "What crater is that just going off?"

Anders looked down and spotted what Carr had seen. "That's some small-impact crater. We will call it John Aaron's," he declared, referring to the twenty-five-year-old Mission Control wunderkind who oversaw the spacecraft's environmental control system and impressed nearly everyone who worked with him with his native smarts.

Aaron, sitting at his console, popped his head up and smiled— and Anders seemed almost to know it. "If he'll keep looking at our systems, anyway," the astronaut said.

"He just quit looking," Carr noted with a wink at Aaron, who promptly went back to work.

☆ ☆ ☆

During the third blackout. Apollo 8's crew was due to conduct the eleven-second circularizing burn. Converting the orbit's shape from an ellipse to a circle had nothing to do with neatness. Half the point of the mission was to study the gravity-distorting mascons buried beneath the surface of the moon, and the ship's current egg-shaped orbit would make it nearly impossible to perform this task well. What was more, a noncritical firing of the SPS would serve as another test of the engine's ability to conduct the decidedly more important burn after the tenth orbit.

Now feeling confident that their SPS was reliable, the astronauts approached and executed the brief firing almost casually, and after the third blackout ended, they reported that their ship's orbit was indeed a tidy 69 statute miles all around. The third pass over the lunar near side was routine, or as routine as it could be for a crew that had only recently begun circling another world.

By the time the ship slipped into its next blackout, the astronauts were clearly exhausted. Their mission ran on Houston time to spare their body clocks too much confusion, but that would do them little good if they weren't maintaining regular sleep cycles, and they weren't. All three men had been awake throughout the entire final approach to the moon and the harrowing LOI burn. Now, with seven more revolutions to come and the TEI burn less than fifteen hours away, there was little chance that they would be able to relax and get some sleep before the homeward cruise.

Anders, perhaps more than the other two, had a very crowded schedule. He was barely a quarter of the way through his photography checklist, and toward the end of the fourth blackout, he needed Borman to maneuver the ship in such a way that he would be able to bring the proper landmarks into focus.

"Okay, target twenty-three, frame one twenty-two," Anders said, letting Borman know where the spacecraft needed to be pointed.

Borman gripped his thruster handle and pitched the ship up slightly. He yawned—loudly enough for the cockpit tape to capture the sound—and when the spacecraft was in position, he took a moment to enjoy both the silence in his headset and the new view outside his window.

The spacecraft was now positioned so that, for the first time, the astronauts could see all the way to the horizon of the moon, beyond which lay a huge swath of black sky. None of the men had thought about what that would mean—that when they came around the near side of the moon and once again established a straight line of radio communication between them and the Earth, they would also establish a straight line of sight. That, in turn, would allow for a view of their home planet rising above the bleak lunar plains.

"Oh, my God!" Borman suddenly said. "Look at that picture over there! Here's the Earth coming up. Wow, is that pretty!"

The other two men looked out their windows. Just as Borman said, the blue-white ball that was home to everything they knew—home to every creature and thing and event that had occurred or existed across the entire expanse of the Earth's history—was hovering over the pitted wreck that was the lunar landscape. The astronauts had seen the Earth and they had seen the moon, but this was the first time they were seeing them together—the ugly, broken world beneath them and the lovely, breakable one in front of them.

It was Anders who shook himself from his reverie first, struggling to remove the black-and-white magazine from his camera and replace it with something that would better capture what he was looking at. "Hand me that roll of color, quick," he said to Lovell.

"Oh man, that's great," Lovell said, his eyes still fixed on the apparition outside his window.

"Hurry! Quick!" Anders barked.

Lovell pulled himself away and dove under the seats to the equipment bay. "It's down here?" he asked.

"Just grab me a color," Anders said. "Hurry!"

"Yes, I'm looking for one," Lovell said. "C368?" he asked, reading from a canister.

"Anything, quick!" Anders said.

Lovell reappeared and handed Anders the film, but by then the Earth had drifted out of view.

"Well, I think we missed it," said Anders, his disappointment evident.

"Hey, I got it right here!" Borman exclaimed, having shifted to a different window.

Anders shoved off from the bulkhead and darted to where Borman was positioned. Holding up his camera, he immediately began shooting.

"Take several of them," Borman ordered.

"Take several of them," Lovell echoed.

Anders did. But only one of the images would matter. Only one, now hidden invisibly inside Anders's camera, would eventually move people to understand that worlds—like glass—do break and that the particular world in the photograph needed to be cared for more gently than humans had ever treated it before. That was the picture—the one that would be called *Earthrise*—that rested inside Bill Anders's camera.

But on Christmas Eve day 1968, nobody knew it.

FOURTEEN

Christmas Eve 1968

THE TELEVISION NETWORKS WOULD GET THEIR BILLION PEOPLE.
They didn't think such a thing was possible—or at least not possible yet,
not in 1968, when a billion-person audience would mean that nearly a
third of the people on the planet had stopped what they were doing to
turn on a television set and watch the same broadcast at the same
moment.

But the numbers coming in to the networks said otherwise. It wasn't
just North America and Western Europe that watched the first, poorly
timed broadcast from Apollo 8 in lunar orbit. It was all of Europe, on
both sides of the East-West divide—including the grim, gray bunker that
was East Berlin—with the nations in the Soviet bloc either parting the
Iron Curtain to let the broadcast through or giving up the fight to prevent the people who lived there from pirating the television signal. That
first broadcast was watched as well in Central America and South America and Japan and South Korea and throughout much of the war zone
that was Southeast Asia. It was watched in India and Africa and Austra-
lia. It was watched on American naval vessels and military bases around
the world. It was watched more or less anywhere there were TV sets,

electricity, and human beings interested in tuning in to the first-ever broadcast from a different world.

"Your TV program was a big success," Mike Collins called up to the spacecraft late on the morning of Christmas Eve. "It was viewed by most of the nations of your neighboring planet, the Earth."

That audience would surely be dwarfed by the one for the evening broadcast that was still to come—the show that would air during the Western Hemisphere's Christmas Eve, when people would be gathered with their families and friends and, in the American time zones at least, would have just had their dinners and sung their holiday songs and settled back on their sofas to hear what the men circling the moon had to say to them. The three astronauts would be on their ninth lunar revolution when they spoke, and after the tenth they would attempt to fire their engine again and return to Earth. If the engine worked, the broadcast would be remembered as a lyric celebration of a job brilliantly done. If it did not work, the broadcast would be known as an elegy.

The NASA image makers may not have fully anticipated just how huge the viewership for the television show would be, but they'd suspected that it would set records. And even before the spacecraft had left the Earth, they had decided that whatever the crew said, it had better be good.

"You're going to have the largest audience that's ever listened to or seen a television picture of a human being, on Christmas Eve," Julian Scheer, the chief public affairs officer in Houston, said when he first raised the matter to Borman, weeks before liftoff. "And you've got, I don't know, five or six minutes."

"Well, that's great, Julian," Borman answered. "What are we doing?"

"Do whatever's appropriate" was all Scheer could suggest.

Borman chewed that over for a while but couldn't come up with much. Then he discussed it with Lovell and Anders, but they were stumped, too. So Borman approached Si Bourgin, a friend who worked for the U.S. Information Agency, to see what he might suggest. But Bourgin, whose job was to frame the American story in the most appealing way possible for the consumption of the world, drew a blank as well. He then farmed the problem out to Joe Laitin, a former wire service reporter who had later become a public affairs officer for President Kennedy and currently worked for President Johnson.

Laitin agreed to give the matter some thought. He tried out a few notions, got nowhere, and then asked his wife, Christine, if she had any ideas about what the astronauts should say; she, too, could suggest nothing. One night, he sat in his kitchen and typed out what he hoped were the appropriate words; upon rereading them, he balled up the piece of paper and started over. Hours went by, and more pages came out of his typewriter only to be tossed aside. Finally, at four in the morning, his wife padded into the kitchen and said she had an idea. She described it and he listened and he smiled. It was perfect.

Laitin passed the idea to Borman, Lovell, and Anders; they approved it, as did NASA. So the words Joe Laitin's wife had suggested were then typed onto a piece of fireproof paper—since Apollo 1, the only kind of paper allowed in a spacecraft—and the page was inserted at the back of the flight plan. There it would remain until Christmas Eve, when the mission to the moon would be nearly done.

☆ ☆ ☆

The day unspooled slowly in Houston—too slowly for the NASA employees who were involved in the mission, to say nothing of the much smaller group of people who lived with and loved the men flying it. The mission controllers who weren't on duty idled about their homes until they couldn't take it anymore, at which point they drove to the space center. Many of them appeared hours before their shifts were due to begin; their early arrival wasn't discussed or planned, and it didn't need to be.

Since the beginning of the space program, Mission Control had operated on a three-shift cycle during flights. But NASA was already planning to expand it to four: eight hours on and sixteen hours off was simply too tiring for console teams doing the white-knuckle work of sending men to the moon. If the Apollo 8 controllers were showing the strain, however, today none of that mattered. Chris Kraft, Gene Kranz, Bob Gilruth, George Mueller, and the other members of the space agency brass barely left the Mission Control auditorium except to go home for a shower, a change of clothes, and, if absolutely necessary, a brief nap.

As the day went on and the other controllers arrived, they looked for a free chair and dragged it over to the console where they usually worked, sitting behind the shoulder of the man on duty. The controllers who couldn't find a chair simply stood against the wall or wandered about the room.

When Milt Windler walked in, he was carrying a bundle filled with his red-and-blue flags with their white numeral 1. Assuming all went well, he would hand out the flags whenever the moment seemed right.

Late in the afternoon, when Mike Collins finished his shift, Borman made a point of calling down to the new capcom, Ken Mattingly, who was one of the space program's few unmarried astronauts. "Hey, Ken," Borman said. "How'd you pull a duty on Christmas Eve? It happens to bachelors all the time, doesn't it?"

"I wouldn't be anywhere else tonight," Mattingly answered.

In the Borman, Lovell, and Anders households, the wives worked on as they had for much of the week, which meant managing the children and hosting the guests and satisfying the unceasing needs of the media. Susan Borman had slept little and felt ragged; she feared she looked that way, too. Days earlier, she had scheduled a beauty parlor appointment for today, partly because she knew she would have hours to fill and might as well look nice for the photographers and cameramen and partly because that was exactly the kind of business-as-usual behavior that NASA liked the world to see.

Valerie Anders remained indoors with her children and came outside to talk whenever the press needed a statement; otherwise, she remained unseen. As the rookie's wife, Valerie was expected to have a rookie's nerves. But she didn't, mainly because she was good at managing her mood by managing her environment.

Like Susan Borman and Marilyn Lovell, she'd known how critical the LOI burn was, and she had stayed awake for it. But she'd passed those crucial minutes in her own way, listening to the squawk box in her darkened living room, with the Christmas tree and the fireplace providing the only illumination. She was surrounded by a tight clutch of people that included an old school friend as well as Jan Armstrong—the wife of astronaut Neil Armstrong—and Dave Scott, Armstrong's copilot on Gemini 8 and one of the three astronauts who would fly Apollo 9 just two months later. As the burn played out, Scott leaned in toward the squawk box and explained any of the gibberish she wasn't able to follow.

When Valerie stepped outside to address the reporters after learning that the burn had been successful, she did so with a polish suited more to a senator than to the young wife of a hero.

"The significance of historical events cannot be realized immediately," she said, "nor can the impact or magnitude of the event be adequately described at the time of the occurrence. Though history is being made today, we all need to try and comprehend the years of effort by many people involved in the eventual lunar landing."

The reporters raced to take it all down. Valerie, her job complete, went back inside to make sure her five young children hadn't wrecked the house in the few short moments she had been gone.

Marilyn Lovell—the most experienced of the three wives and, indeed, the most experienced astronaut's wife in the world, measured by the amount of time her husband had spent in space—knew she would have to make time for a nap, and she did so, too, just as soon as the LOI burn was finished and the first broadcast had signed off. She also knew that if she could possibly manage it, she needed a few minutes out of the house.

Christmas week had always meant church for her and her family, and they had found a comfortable Episcopal congregation in Houston with a popular minister, Father Donald Raish, who knew all of the Lovells well. Attending a proper Christmas Eve service today would be impossible, what with the business of getting the children ready and navigating the media storm that would erupt the moment the entire family emerged on the front lawn.

Instead, Marilyn called Father Raish and asked if she might come by early, perhaps to have a private moment inside the church. He agreed, and late in the afternoon she left the house alone, fielded as few questions from the press as she could, hopped in her car, and drove to the church.

When she arrived and went inside, she could see that the sanctuary was almost empty, save for Father Raish and the organist. Candles were lit, the organ was playing, and Father Raish greeted her. They walked to the altar, and Marilyn prayed. She was happy for the quiet moment and grateful for the effort that had been made on her behalf.

"You did all this for me?" she asked Father Raish.

"Well, you have to miss Mass tonight, so yes," he answered modestly.

Marilyn smiled, thanked him, and hurried back out to her car. It was already getting dark, but avoiding the crowds meant that she would have to take the poorly lit back roads into the Timber Cove neighborhood. After so many years, however, she knew the route well. As she approached her home, the trees above the road appeared to part: directly in front of her, the moon hung in the sky. She stopped the car and looked up at it.

Jim is up there, she thought.

Marilyn sat with that idea for a moment, then drove the rest of the way to her brightly lit and very crowded home. The nighttime broadcast was still hours away, and the house would only get brighter and more crowded still.

☆ ☆ ☆

Of the three men circling the moon in the small Apollo spacecraft, Borman was the only one who would admit that he was flat-out exhausted. Lovell seemed tireless, but then Lovell was never as happy as when he was in space. He also loved being busy, and the complex navigation of the translunar route meant he almost always had a lot to do. Even when the current flight plan called for a rest period, he was perfectly content to vanish into the equipment bay to take his sightings and make fine corrections in the trajectory.

Anders, meanwhile, could not sit still. He was five years younger than Borman and Lovell, but the way he bounced from window to window, taking his sightings and snapping his pictures, put Borman in mind of his boys when they'd been in grade school.

In truth, however, both Anders and Lovell were wrung out; Borman could see it in the red that rimmed their eyes and the periodic yawns they tried to stifle. What was more, while Lovell would sometimes boast, with justification, about how deftly he could play the computer's keypad—"like a concert pianist," he liked to say—he couldn't hide his mistakes. Mostly the computer was silent, but when Lovell entered an erroneous command, the system would emit a warning tone. Borman had heard a few too many of those alerts, another sign that fatigue was taking a toll.

Although mission rules allowed all three astronauts to schedule their

sleep interval for the same time, Borman had made it clear that he preferred it if one of them kept watch.

"I'm going to sack out for an hour," he told Lovell and Anders. "One
of you should, too."

Lovell, in the equipment bay, nodded vaguely. Anders held up a
hand in a "just a minute" gesture.

Borman did not care to pull rank—not yet, anyway—but he could at
least set a good example. He floated down to the equipment bay, rigged
his sleeping bag, and floated back out to confirm that his ship was in
order. After deciding that it was, he drifted back down to the equipment
bay.

Moments later, Anders and Lovell could hear Borman breathing
steadily. For the next hour, the two astronauts worked together to help
Anders catch up on his photography. Lovell—whose formal title was
command module pilot and who had both the ability and the authority
to take the wheel of the ship whenever necessary—maneuvered the
spacecraft to give Anders the angles he needed. Anders changed windows as required, working to get the alignment for each picture right.

Toward the end of the hour, the two men began struggling with a
particularly difficult target.

"Is it ten degrees left?" Lovell whispered.

"Yes, right now. You got enough roll?" Anders asked.

"I'll get it for you."

"That you driving it?"

"I'm driving it now, yes."

They continued whispering back and forth as the spacecraft moved
about. At one point, Lovell overshot his mark and fired a counterthrust
to halt the ship.

"Ahhh!" he exclaimed.

"What happened?" Anders asked.

"It's okay," Lovell reassured him.

"Don't wake up Frank."

But it was too late; Borman was awake. The motion of the ship and
the ongoing background chatter had proved too much. He drifted out of
his sleeping bag, rubbed his eyes, and looked around.

"Sorry, Frank, didn't mean to disturb you," Lovell said.

Borman waved it off. The hour of sleep had helped; though he didn't feel anything like fully rested, he did feel much improved. Looking at his crewmen, however, he could see that the same hour hadn't done them any favors; the fatigue showing in their faces was becoming ever more evident. He would give them a little more time, through the remainder of the seventh rev, and then whether they liked it or not, they'd go belowdecks, climb into their sleeping bags, and, as they put it at West Point, go "local horizontal."

Borman gave them that time, and they all spent it working through their checklists and the other chores the capcom radioed up. Then, at the end of the orbit, Collins called with yet another series of tasks: a battery charge, a cryogenics stir, a telemetry check, all in addition to the ongoing work involving the surface sightings.

That was it. Borman keyed open his mike in annoyance.

"Houston, Apollo 8," he called down.

"Roger, go ahead, Frank."

"I want to scrub these control point sightings on this next rev," he declared.

"I understand you want to scrub control points one, two, three on the next rev," Collins said.

"That's right. We're all getting too tired."

"Okay, Frank."

Then, just to be clear, Borman keyed his mike open again. "We're scrubbing everything," he declared. "I'll stay up and keep the spacecraft vertical and take some automatic pictures. But I want Bill and Jim to get some sleep."

"Roger," Collins responded. "Understood."

"Unbelievable, the details these guys send," Borman muttered. "Just completely unrealistic stuff."

"I'm willing to try it," Anders suggested tentatively, indicating his cameras and his flight plan.

"No, you try it and we'll make another mistake," Borman said. "I want you to get your ass in bed." When Anders hesitated, Borman practically barked at him: "Right now! Get to bed! I'm not kidding you."

"Shall we . . . shall we do that thing over there?" Anders asked, gesturing vaguely to a target somewhere out the window that he had been preparing to shoot.

Borman rolled his eyes. "This is a closed issue," he said. "Go to sleep, both of you guys." Anders, still dithering with his cameras, stalled some more. "To hell with the other stuff!" Borman said sharply. "You should see your eyes; get to bed."

Finally relenting, Anders slumped off to the equipment bay.

"A quick snooze and you guys will feel a hell of a lot better," Borman said, his tone a little softer now.

Lovell, unlike Anders, hadn't needed any persuading. He had flown under Borman's command before—and chafed under it, too. But this time he was more than willing to submit to Borman's authority. Without having to be asked a second time, he drifted down to his own sleeping area and was out almost instantly.

Borman settled back into his left-hand seat. "Lovell's snoring already," he radioed quietly to the ground.

"Yes," Collins answered, "we can hear him down here."

☆ ☆ ☆

As the time for the Christmas Eve television show approached, Borman, Lovell, and Anders were flying over the far side of the moon. They had completed their seventh orbit—with Lovell and Anders sound asleep through most of it—and then their eighth. Now they had barely fifteen minutes left until reacquisition of signal, when the broadcast would begin, and they were still not entirely clear about how they were going to fill the time.

Julian Scheer had originally promised that the show would last just six or seven minutes; the flight plan, however, called for at least twenty. Yes, the last couple of minutes were planned, but as for the rest, the astronauts had been so busy with other things that they hadn't thought much about how they would inform and entertain their enormous audience.

"We've got to do it up right," Borman said as Anders unstowed the camera and began fooling with focal angles. Then, just in case his fellow crew members had forgotten, he added, "There will be more people listening to this than ever listened to any single person in history."

Neither Lovell nor Anders offered any useful ideas about how the three of them should amuse all of those people, so Borman made a suggestion. "Why don't we each talk about one thing that impressed us most out of what we saw and describe it," he said.

There was silence.

"Okay?" Borman asked.

The other two shrugged, and the matter was settled.

As the crew sailed through the final miles of their far-side orbit, Walter Cronkite went live from his studio in New York. The astronauts might find it difficult to fill their airtime, but Cronkite, with a lifetime of on-camera experience, had no trouble keeping the chatter going until the signal from the spacecraft was acquired.

"Apollo 8 is in its ninth and next-to-last full orbit of the moon," he began. "The astronauts, on the orders of command pilot Frank Borman, are scrubbing all remaining items from their flight plan except one more television transmission, which should come up very shortly now, because they are tired and need to rest before the critical maneuver that starts them back to Earth early tomorrow morning."

In Houston, Mission Control reported that it had acquired the signal from the spacecraft. But the transmission was initially scratchy, and a full two minutes would elapse before voice and pictures could accompany the basic telemetry.

That was just as well, because the ship's crew clearly wasn't ready to go on the air. Borman was at his couch, working the attitude controller and struggling to bring the lunar surface into frame. Anders, fighting to get the high-gain antenna locked onto the Earth, adjusted the communications systems, which were on his side of the instrument panel.

"You'll have to pitch up, Frank," he said. "Pitch up or yaw right."

"I'll do both," Borman replied.

As he maneuvered the ship with one hand, Borman picked up the TV camera with the other, hoping he could capture a good angle. A second later, the surface of the moon slid into the viewfinder in sharp focus.

"Here it comes!" he said.

"Okay!" Anders answered.

"Oh boy!" Borman exclaimed.

"Got a good shot of her?" Lovell asked.

"Yes!"

"Well, keep the camera there!" Lovell admonished. "Keep the camera."

On the ground, Cronkite talked on: "We're all anxiously looking forward to this second set of pictures of the moon from seventy miles high and the spacecraft moving across the moon's surface at thirty-six hundred miles per hour." Numbers and names, the anchorman knew, were a good way to add substance to any part of a broadcast that was really just an exercise in filling time. "The person you'll be hearing speaking to Apollo 8 is astronaut Ken Mattingly, who is so-called capcom, capsule communicator."

Finally, a voice-and-picture link was established. The screens in Mission Control, the New York studios, and a billion homes around the world were once again filled with the moon.

"How's the TV look, Houston?" Anders asked.

"Loud and clear," Mattingly answered.

"It looks okay?"

"Very good."

Satisfied, Borman began: "This is Apollo 8, coming to you live from the moon." His voice, traveling across a quarter million miles of void, sounded thinner and more nasal than it really was. But it was steady and strong, and so was the signal that carried it. "Bill Anders, Jim Lovell, and myself have spent the day before Christmas doing experiments, taking pictures, and firing our spacecraft engines to maneuver around. What we will do now is follow the trail we've been following all day and take you on through a lunar sunset."

The static crackled but the picture stayed true, and Borman went on, following the narrative the crew had planned. "The moon is a different thing to each one of us," he said. "I know my own impression is that it's a vast, lonely, forbidding-type existence or expanse of nothing. It looks like clouds and clouds of pumice stone, and certainly not a very inviting place to live or work. Jim, what have you thought most about?"

"My thoughts are very similar," Lovell said. "The vast loneliness up here of the moon is awe-inspiring, and it makes you realize just what you have back there on Earth. The Earth from here is a grand oasis in the big vastness of space."

"Bill, what do you think?" Borman asked.

"I think the one thing that impressed me most was the lunar sunrises and sunsets," said Anders, still very much the photographer. "These in particular bring out the stark nature of the terrain, and the long shadows really bring out the relief that is hard to see at this very bright surface that we're going over now."

They were trying, these pilots who had been asked to play poets, and they were performing reasonably well. But it was Anders, staying true to what he was—a professional on a mission—who sounded the most authentic, and Borman and Lovell knew it. So for the rest of the broadcast, they made an unspoken decision to play it straight and simply narrate what they were seeing.

"We are now coming onto Smyth's Sea, a small mare region covered with a dark level material," Anders said.

"What you're seeing across Smyth's Sea are the craters Kästner and Gilbert," Lovell said.

"The horizon here is very, very stark," Anders said. "The sky is pitch-black, and the Earth"—Anders caught himself—"or the moon, rather, excuse me, is quite light." Moon below, Earth in the sky; it still took getting used to.

The audience encountered the Marsh of Sleep again, and the Seas of Tranquillity, Fertility, and Crises. And then, finally, the ship approached what was known, too ominously, as the lunar terminator—the clean, sharp line on the airless moon where daylight instantly gave way to nighttime, with none of the luminous wash of sun through atmosphere that makes dawn and dusk so gradual on Earth. The thing about the terminator, though, was that it was two things at once, either the dying of the light or the arrival of it, depending on whether you were standing in the shadow and looking at the sun or standing in the sun and looking at the shadow. Borman knew which one he preferred.

"Now you can see the long shadows of the lunar sunrise," he said, "and for all the people back on Earth, the crew of Apollo 8 has a message we would like to send you."

Anders picked up the flight plan and turned to the last page. Poetry might be alien to the three men, but at last they had verse to speak.

"In the beginning," Anders began, "God created the heaven and the

Earth. And the Earth was without form and void, and darkness was upon the face of the deep; and the spirit of God moved upon the face of the waters. And God said, 'Let there be light,' and there was light."

Anders read a bit more, and then Lovell picked up the words. "And God called the light Day and the darkness He called Night. And the evening and the morning were the first day. And God said, 'Let there be a firmament in the midst of the waters. And let it divide the waters from the waters.'"

Finally, when Lovell was through, Borman finished up. He read the ancient passage about the gathering of the waters, the appearance of the dry land, and the naming of the land and the waters, concluding with the words, "And God saw that it was good." Then the commander of the crew that had ventured so far and seen so much spoke to the billion or so people who represented the entire questing species for whom they had made their journey.

"And from the crew of Apollo 8, we close with good night, good luck, a merry Christmas, and God bless all of you, all of you on the good Earth."

Then the signal from the spacecraft cut off and the men circling the moon were once again alone, beings apart. Now, with the broadcast over, it would be up to the men and women and children on the home planet to make what they would of the images they had just seen and the words they had just heard.

☆ ☆ ☆

Frank Borman, finished with the television broadcast, was still in command of an active spacecraft, and that spacecraft was still locked in a repeating orbit around the moon. So once the TV show was over, he was eager to get back to the business at hand.

On the ground, many of the controllers were not quite ready to let go of the feelings they had just experienced. Gene Kranz, for all his military grit, was also something of a sentimentalist, which he would readily admit. When the broadcast from Apollo 8 ended, he stood quietly at his back-of-the-room console in Mission Control, still filled with the rhapsody of what he had just seen.

Jerry Bostick, the flight dynamics officer at his console in the trench, felt something he could describe only as a wave of gratitude—for the astonishing moment in history that was unfolding in front of him, and for the accident of birth and timing and talent that had placed him, one person out of billions, in the middle of that moment. *Thank you, Lord, for letting me be here and be a part of this,* he said to himself silently.

Milt Windler, now sitting at the flight director's console, considered his bundle of happy flags with their big white numeral and decided that they would forever remain packed away. The secular cathedral of Mission Control had, if only briefly, become a spiritual one. A pep rally, no matter how well intentioned, suddenly seemed entirely out of place.

Borman, meanwhile, waited until he was reasonably sure the broadcast had ended and then waited a beat longer. Sensing the power of the moment that had just passed, he wanted to make certain that his next words would get through to Mission Control but not go beyond.

"Are we off the air now?" he asked.

"That's affirmative, Apollo 8, you are," Ken Mattingly answered evenly.

"Did you read everything we had to say there?"

"Loud and clear. Thank you for a real good show."

"Okay. Now, Ken, we'd like to get squared away for TEI here," Borman said. "Can you give us some good words like you promised?"

"Yes, sir," Mattingly answered.

The good words Borman wanted were the PADs—the preliminary advisory data, or the computer commands—for the return-to-Earth burn, as well as other details like the navigational star sightings Lovell would need to take and the precise length of the burn. In the simulator, all of the variables in a TEI could be controlled, but in a real burn, any number of factors could be introduced, from orbital wobbles caused by the mascons to trajectory shallowing caused by wastewater dumps. Especially important was the question of precisely how much fuel the SPS had used in its previous burns. The engine ingested 547 pounds of fuel every second it fired, which was a lot of weight to give or take when you were aiming to place a spacecraft in a very narrow reentry corridor in the Earth's atmosphere 233,000 miles away.

Borman was well aware of the importance of these kinds of calculations. During the broadcast, he might have sounded nonchalant when speaking about how the crew had been "firing our spacecraft engines to maneuver around," but that glossed over the fact that "maneuvering around" took a lot of planning and a lot of fuel. Good words from Ken Mattingly would go a long way toward minimizing the attendant risk.

Mattingly spent a fair bit of the ninth revolution sending up the necessary alignments and other settings. Much of it was a delightfully discordant mix of the mathematical and the zodiacal—the strange language of both the ancient mariners who'd once sailed and the new sky mariners flying today.

"Scorpii Delta down 071," Mattingly read off. "Sirius Rigel 129."

Borman and Anders copied it all out in longhand, like operators at telegraph keys. Lovell, at his computer, punched in the coordinates.

"Basically," Mattingly concluded when he was at last done, "all systems are good. After one hundred and thirty-eight seconds of the burn, you are on your way home. The weather in the recovery area looks good."

Borman acknowledged Mattingly's report but made no response to the capcom's mention of the recovery area. The weather could change well before Apollo 8 returned to Earth, and certainly the fortunes of the spacecraft could, too. For now, the crew was headed toward the end of their ninth orbit and their penultimate thirty-five minutes of blackout.

"Thanks a lot," Borman said without ceremony. "We'll see you around the next pass."

☆ ☆ ☆

As the hour of the trans-Earth injection burn approached, Susan Borman decided that she would handle Apollo 8's final burn differently from the way she'd handled the lunar orbit insertion. She still preferred to keep the press and most of the guests in her home at a safe remove during the mission's most critical moments, but the solitary vigil she had maintained for the LOI burn had also left her spent. This time around, she chose to share the high-wire wait with at least one other person by her side, and she very much wanted that person to be Valerie Anders. Like Susan, Valerie lived in the El Lago community, meaning she was just a few lawns and a short stroll away. Marilyn lived in Timber Cove, which would require her to get in her car and drive through the mobs of reporters and photographers in order to get to the Bormans' home. So Susan had invited Valerie to drop by for the Christmas Eve broadcast and the engine burn, and Valerie had accepted.

Susan and her sons had their Christmas Eve dinner at the nearby home of Joe and Margaret Elkins, the Bormans' closest friends in the neighborhood, and then she and the boys hurried back to their house. Valerie put her children to bed, left them in the care of one of the many adults visiting her, and walked over the Bormans'. Before long, the two women slipped away from the crowd of well-wishers in the living room, repaired to the kitchen nook, and began their vigil.

Marilyn had chosen not to change a thing: she would follow the coverage in her home just as she had all of the other critical moments in the mission, with her friends and family grouped around her, in front of the TV. After the crew's broadcast but before the burn, she had taken her children for a walk around the Timber Cove neighborhood, wearing a long red skirt that seemed fitting for Christmas Eve. The nearby houses were always decorated during the holiday season, but this evening Marilyn was surprised to see that almost all of her neighbors had gone further than usual and lined the sidewalks with homemade luminaries. Up and down the street, carefully positioned candles flickered inside paper bags in tribute to the men aboard Apollo 8. Though the display was fragile—a brisk wind or a tipped candle could destroy it—for now the lights were holding.

Marilyn could not personally thank every one of her neighbors, but she could express her appreciation for the attention her family had been shown all week in another way. After she got home and put the children to bed, she went into her kitchen, arranged cups of eggnog on a large tray, and went back to the front lawn to pass the Christmas cheer out to the members of the press, who were very far from their own homes tonight. Then she went back inside and gathered with her guests in front of the TV.

Although Walter Cronkite had signed off the air after the astronauts' broadcast, he had remained in the New York studio, using the brief break to rest and review technical material about the burn. He and the other network anchors then signed back on to cover the final orbit, which was decidedly short on news. The sightseeing was done and the happy narration for the viewers at home was over. What passed for

conversation between ship and shore was now almost entirely the stream
of numbers and technical language that would further refine the systems for the burn. But Cronkite framed the moment in a starker, more
binary way.

"While Apollo 8 is behind the moon, out of touch with the ground
control again, the astronauts are to start up that big rocket engine that
powered them into lunar orbit," he said. "It must work perfectly again,
because if it fails, Apollo 8 could be caught in that lunar orbit. That, of
course, is not expected to happen. The engine has worked perfectly so
far. It should work once more."

That was true as far as it went, but it didn't go far. As the astronauts,
their wives, and everyone at NASA knew very well, completely reliable
systems worked only until the moment they became completely unreli-
able.

☆ ☆ ☆

Shortly before the blackout at the end of the tenth orbit, Ken Mattingly
made the call that began the countdown to the burn.

"Okay, Apollo 8," he said. "We've reviewed all your systems. You
have a go for TEI."

"Okay," Borman answered.

As the clock ticked toward what everyone hoped would be the final
loss of signal, Mattingly began the ritual of calling off the time.

"Apollo 8, Houston. We have three minutes to LOS. All systems are
go."

There was no answer from the ship.

"Apollo 8, Apollo 8, this is Houston," Mattingly repeated. "Three
minutes LOS. All systems are go. Over."

"Roger, thank you, Houston. Apollo 8," Borman responded, with a
clear emphasis on the last two words. There was a conclusory note to
the sign-off, one that did not invite further chatter unless it was necessary.

Mattingly, fluent in the commander's tongue, held his own for the
next three minutes. For the only time in the mission thus far, there would
be no steady countdown to a pivotal event.

Finally, just seconds before blackout, the capcom sent up an all-business good-bye.

"All systems are go, Apollo 8."

"Thank you," Borman answered.

Then the line went dead.

☆ ☆ ☆

Inside the spacecraft, the astronauts took a moment to sit quietly. It was a relief to be freed from the stream of must-do chores that were forever being read off to them. Then they went about making the final preparations for the TEI burn as wordlessly as they could—saying what they had to, avoiding anything else.

When the burn was less than thirty minutes away, Borman finally spoke up again.

"It's been a pretty fantastic week, hasn't it?" he said.

Lovell smiled, thinking of home. "It's going to get better," he said.

For the next twenty minutes, they worked on various routine tasks, including the orientation of their high-gain antenna. Though useless now, it would go back to work when they came around the moon's near side.

Anders glanced out his window. "Boy, it's blacker than pitch out there," he said.

A few more minutes elapsed. Borman looked at his clock, which indicated the time remaining until the burn. "Seven minutes," he said. "Coming up on six minutes."

As they had before, the astronauts settled back into their couches and loosely fastened their seat restraints. This time the spacecraft would be accelerating by 3,552 feet per second, or 2,422 miles per hour, up to a peak speed of at least 5,324 miles per hour. That would be a lot of acceleration in just over two minutes. Once again the crew would be pushed back in their seats with a comparatively gentle hand of less than one g, though after four days of no g's at all, it would feel like a lot more.

Anders called more attitude coordinates from the flight plan. In response, Borman fine-tuned the ship in the pitch and yaw axes.

"Stand by for two minutes," Anders said.

After the final seconds ticked down, the computer flashed its 99:20 last-chance code.

Lovell reached forward and pressed PROCEED.

☆ ☆ ☆

The atmosphere in Mission Control wasn't at all what Chris Kraft wanted it to be. His controllers were professionals, and they generally abided by the rules of the room, which meant concentrating on the work and keeping extraneous talk to a minimum. But those rules were usually relaxed during blackouts, and though the mood had been tense during Apollo 8's first loss of signal, the experience was familiar by now. With the headsets silent and the telemetry shooting blanks, why should anyone object if a man exchanged a few words with the fellow to his left or right?

This blackout, however, was different. In less than twenty minutes, Mission Control would get either very good news or very bad news, depending on what the SPS did. If the news was bad, it would likely mean the loss of three good men. Kraft now knew, as he hadn't known twenty-three months before, what losing three men felt like, and he never wanted to experience such a thing again. So although the room was reasonably quiet, under the circumstances it was not as quiet as Kraft felt it should be.

"Could you please shut up over there?" he barked.

Heads swiveled in response.

Kraft wasn't directing the remark at any one person in particular. He heard a low hum of talk off to one side, and if he had to pick one person whose voice was irritating him the most, he guessed it would be that of Charles Berry, the flight surgeon. Babysitting the astronauts' health was necessary, of course, but at the moment it was entirely beside the point. You couldn't prescribe Marezine or Seconal for a busted engine, and in Kraft's view, it would be nice if Berry, of all people, stayed as far out of the way as possible tonight.

"Could you please shut up?" Kraft repeated. "I don't know about the rest of you, but I'm thinking about whether these guys are going to come out the other side."

Kraft looked away, scanning the room and darkening his expression to make it clear that the scolding was for everyone.

In El Lago and Timber Cove, nobody had to ask for silence. Susan Borman and Valerie Anders were seated side by side, alone, in the Bormans' kitchen nook. Marilyn Lovell sat on her living room floor with her knees tucked up, just as she had during the television broadcast hours earlier.

On the bottom of the TV screens tuned to the coverage of Apollo 8 and at the top of the main screen in Mission Control, the mission's elapsed time ticked up to three days, seventeen hours, nineteen minutes, and twelve seconds. Some of the viewers at home might have known without being told that this was the moment the burn was supposed to begin. Every single person in Mission Control knew it well. If the engine lit as planned, it would speed the ship to an early exit from its final blackout, and the crew would regain radio contact just fifteen minutes and ten seconds later.

The engine burn—invisibly, unknowably—either did or didn't take place, and the long wait for news from the ship passed the way Kraft wanted it to, in silence. He said nothing. George Low said nothing. Charles Berry said nothing. Ken Mattingly, at the capcom console, did not mark the start and stop times of the burn when they arrived and passed; if you couldn't confirm that an event had happened, you didn't announce it. Instead, fingers were drummed and Zippos were rasped and the mission clock counted off the time.

A bit less than twelve minutes after the burn should have happened, Mattingly made his first call. As with the orbital insertion burn, it was too soon to expect reacquisition of signal, but not too soon to begin hoping for it.

"Apollo 8, Houston," Mattingly said to the invisible spacecraft and, by extension, to the global television audience.

He was met by silence.

"Apollo 8, Houston," he repeated eighteen seconds later.

Again nothing—only the loud hiss of cosmic absence in his headset and on the TVs.

He tried again after twenty-eight more seconds. "Apollo 8, Houston."

This time, hearing nothing, he gave it nearly a minute. His next call received no response, either.

After letting forty-eight more seconds pass, Mattingly hailed the ship again, then let yet another forty-eight seconds pass and tried once more. More silence.

Now the fifteen minutes and ten seconds were well and truly gone. If the spacecraft did not appear soon, it would be seriously overdue.

And then, as one, the dead screens at every console in front of every man in Mission Control began to jump and flicker. All at once they were filled with numbers—beautifully complete and beautifully healthy numbers—streaming to them from a spacecraft that was still a quarter million miles away but was undeniably speeding toward home.

A moment later, Jim Lovell's voice came through clearly.

"Houston, Apollo 8," he said. "Please be informed there is a Santa Claus."

"That's affirmative," Mattingly responded. "You are the best ones to know."

That, at least, is what the astronauts heard. But the people in Mission Control did not, because the capcom's words were drowned out by the whooping and cheering and whistling from the controllers who now didn't give a fig for Chris Kraft's rule of silence.

In the Bormans' kitchen nook, Susan leapt up, waved her hands excitedly in the air, and then brought them down to take Valerie in a happy hug. In the Lovell home, Marilyn stood at the center of a similarly raucous scene, and the noisy crowd soon woke the children she had just put to sleep.

There was much applause, too, in homes and in bars and on American military bases around the world. Also celebrating the moment were the seventeen hundred men aboard the aircraft carrier USS *Yorktown*, the prime recovery vessel, which was already steaming into position in the South Pacific for the splashdown that was now just over two days away.

From the spacecraft, which still had a long way to go before it landed in that small spot of ocean, Frank Borman had only one question.

"What's next on the docket?" he asked.

FIFTEEN

December 25–27, 1968

D EKE SLAYTON HAD NOT SPOKEN DIRECTLY TO A CREW IN SPACE since October, when he'd made his futile attempt to persuade Wally Schirra to put his helmet on before Apollo 7 reentered the atmosphere. Slayton was well aware that communications protocols frowned on any-one but the capcom—even the head of the astronaut office—talking to the crews, but after the TEI burn, he felt like making another exception.

"Good morning, Apollo 8, Deke here," he said, after plugging his headset into a second jack at the capcom station. "I just would like to wish you a very merry Christmas on behalf of everyone in the control center and, I'm sure, everyone around the world."

Slayton getting even a little sappy was not something the crew had ever experienced before, and they did not know quite how to respond. That mattered little, however, because apparently a sappy Slayton was also a garrulous Slayton.

"None of us ever expected to have a better Christmas present than this one. Hope you get a good night's sleep from here on and enjoy your Christmas dinner tomorrow. We look forward to seeing you in Hawaii on the twenty-eighth."

Borman waited till he was sure Slayton was done and only then responded. "Okay, leader," he began, choosing an unexpectedly deferential form of address and one he hadn't used before. But the commander of the mission was not entirely immune to sentiment, either. "We'll see you there," he said. "Thank everyone on the ground for us. It's pretty clear we wouldn't be anywhere if we didn't have them helping us out."

"We concur that," Slayton said.

"Even Mr. Kraft does something right once in a while," Anders added.

Slayton glanced over his shoulder to see if Kraft had taken the ship-to-shore impudence in good humor. But Kraft wasn't in the room; with the spacecraft at last on its way to Earth, he had run home for a nap and a change of clothes.

"He got tired of waiting for you to talk and went home," Slayton answered.

As Slayton hoped they would, the astronauts did get some sleep—Borman and Lovell first and then Anders when they awoke. And they did have their Christmas dinner, later in the day on Christmas proper.

Like all the other meals they'd had so far, these were wrapped and stacked in a bin in the equipment bay. But the Christmas dinner packages were heavier than usual, and each was tied up with a green fireproof ribbon with a card reading, "Merry Christmas." When the astronauts took off the outer wrapping, they discovered that the plastic pouches contained turkey and gravy and cranberry sauce and stuffing. Accompanying the dinner was a miniature bottle of Coronet VSQ brandy for each man.

All three astronauts smiled at the meal, which, after four days of flight rations, looked like a feast. And two of the three men—Lovell and Anders—were glad to see the brandy. Borman wasn't.

"For crying out loud, don't drink that," he said. "If anything at all goes wrong in the next two days, they'll blame the problem on a drunken crew."

Lovell and Anders had no intention of opening the bottles. Though they could both handle their liquor, they were still too fatigued to enjoy the indulgence. What's more, they knew Borman was right;

especially now that they were on their way home, they couldn't afford to get sloppy.

Not long after they ate, Lovell radioed down to the ground. "It appears we did a grave injustice to the food people," he said. "Santa Claus brought us a TV dinner each, which was delicious: turkey and gravy, cranberry sauce, grape punch, outstanding." He stressed the "grape punch" part, just to be safe.

At the flight director's station, Milt Windler laughed. In front of him, on a clear spot on the console's little desk area, a cup of coffee and a baloney sandwich rested on a crinkled piece of waxed paper. Just this once, the men in space were eating better than the men on the ground.

☆ ☆ ☆

The Apollo 8 wives spent Christmas Day in a more traditional way. Valerie Anders gathered up her five children and drove to nearby Ellington Air Force Base, where they attended Catholic services at the base chapel. Susan Borman, accompanied by her boys and her parents-in-law, drove to St. Christopher's Episcopal Church, where the Reverend James Buckner read a prayer particular to the most conspicuous member of the congregation, who could not be in attendance that day:

"O, eternal God, in whose domain are all the planets, stars, and galaxies and all the reaches of time and space, from infinity to infinity," he said, "watch over and protect, we pray, the astronauts of our country."

Marilyn Lovell's Christmas morning began with an act of charity. Almost everyone in the media crowd that had filled her lawn for most of the week had briefly decamped to celebrate the day, and now there was just one photographer left. Marilyn peeked out her window and took pity on the man.

"Why don't you go home?" she asked, standing on the front step with her arms folded against the chill. "We're not coming out now."

"I can't go home," the man said forlornly.

"Why not?" Marilyn asked.

"I can't leave until I get a picture."

Marilyn laughed. "Is that all?" she said.

Looking over her shoulder into the living room, she saw that her youngest, Jeffrey, and her oldest, Barbara, were sitting by the Christmas tree amid the morning's presents.

"Kids!" she called. "Come over here." Both children jumped up, but they were empty-handed. "No, no," she said, gesturing toward the mound of gifts, "bring something."

Barbara collected whatever toy was in reach, Jeffrey grabbed a pogo stick that he'd been trying to master for much of the morning, and they both went outside. Jeffrey bounced while Barbara posed, and the reporter took a few pictures. Then he hurried off to spend a few hours of Christmas Day on his own time.

Not long afterward, Marilyn's holiday generosity was repaid. Unannounced, a Rolls-Royce pulled up to the curb and a uniformed man climbed out. He came up the walk, knocked on the door, and presented Marilyn with a box from the Neiman Marcus department store. It was wrapped in sky blue paper and decorated with two Styrofoam balls—one larger, one smaller; one painted like the Earth, one painted like the moon. Attached was a note that read, "Merry Christmas. With love from the man in the moon." Inside was a mink jacket.

Marilyn Lovell, wife of the moon man, wore her gift to church with her children by her side that morning. And if it looked to anyone in the Houston community as if she was putting on airs, well, let it. She had done more than enough this week to earn the indulgence.

☆ ☆ ☆

By late on Christmas Day, a lot of people, including some in the media, had begun to act as if the Apollo 8 mission was effectively over. Yes, the crew still had to land their spacecraft, but that was really just a formality, wasn't it? Commentators on television had begun offering guesses about which cities—aside from cursed New York with its cursed ticker-tape—would host parades and where the astronauts would go on the world tour that would follow. Especially intense was the speculation about when the surely inevitable lunar landing would take place.

The betting was that it could happen as early as May, when Apollo 10 was due to be launched. After a quick shakedown run for the LEM in Earth orbit, it would be off to stamp some bootprints in the Sea of Tranquillity. The crew would include Tom Stafford, Wally Schirra's copilot when Gemini 6 rendezvoused with Gemini 7 in Earth orbit, and he would probably draw the card as first man on the moon. John Young—whom Gus Grissom had chosen above Borman for Gemini 3—and Gene Cernan would fly with him. The press was already starting to prepare the Stafford profiles.

The cocksure idea that Apollo 8's ride home was little more than an easy free fall got an unintentional boost from Bill Anders when the spacecraft had traveled about 45,000 miles from the moon and crossed over the invisible line at which lunar gravity gave way to the Earth's more powerful gravity. When Collins mentioned on the capcom loop that his young son had asked him which of the astronauts was driving the spacecraft, Anders responded, "I think Isaac Newton is doing most of the driving right now."

Physics alone would not bring the crew home, of course, but even Mission Control began sounding a little overconfident. Not long after Anders's comment, Mattingly gave Borman a read of the mood on Earth and related, as well, a conversation he had just had with Windler. "Everybody is smiling," he said. "Santa was good to most of the folks in the world, and everything is pretty calm, like it should be on Christmas."

"Very good," Borman answered.

"Milt says we're in a state of relaxed vigilance."

"Very good," Borman repeated, before realizing what Mattingly had just said. "We'll relax," he said, his tone a bit stern. "You be vigilant."

"That's a fair trade," Mattingly responded.

Borman had a right to demand that vigilance. Apollo 8's flight back to the home planet might appear to be the equivalent of long-distance skydiving, but as with skydiving, it could still go badly awry. In the hours since the TEI, in fact, the mission had nearly come undone, though the news was little reported on Earth.

Borman had been sleeping off his Christmas dinner in the equipment bay while Anders minded the ship in the left-hand couch and Lovell worked the computer. Collins was reading Lovell new coordinates for the rotisserie roll, which kept the spacecraft evenly heated on all sides. Now and then, the attitude of the ship had to be changed depending on its ever-shifting angle relative to the sun. Lovell was punching the commands into the computer when suddenly the thrusters began firing.

The spacecraft slewed dizzyingly, causing it to swing from its roughly nose-forward position to a straight nose-up attitude.

"Whoa, whoa, whoa!" Lovell exclaimed.

"Okay, whoa, whoa. Standing by," Collins responded.

Anders watched as the attitude indicator on the instrument panel swung in parallel to the ship, providing a disturbing readout of the equally disturbing motion; Borman awoke with a start.

"What happened?" he demanded.

For the moment, Lovell didn't have an answer, and the spacecraft continued to lurch. Anders fired the thrusters to return the ship to its proper orientation, but whatever bit of bad code had initiated the problem was fighting back, working to maintain the nose-up position. Knowing that the first rule for this kind of situation was to avoid introducing new problems while one was already unfolding, Anders let go of the thruster handle.

Lovell ran his computer commands in his head and quickly figured out what he had done. Collins had called up the verb 3723 and the noun 501, which would have positioned the ship at the proper roll angle. Lovell, his fatigue getting the better of him, had inadvertently shortened it to verb 37 and noun 01. Those were very different commands: verb 37 meant "return to Earth," and noun 01 meant "prelaunch mode." In effect, Lovell had told the spacecraft that it was on the launchpad in Florida, and the spacecraft had believed him—hence its proud nose-up position as it prepared for liftoff.

"It was my goof," Lovell said.

Collins, who had already correctly guessed what the cause of the problem was, answered simply, "Roger."

The scrambled coding could be fixed, but it would take some work. Because of the incorrect commands, the spacecraft's brain had been wiped clean of any knowledge of its current orientation. Restoring it would require Lovell to take what was known as a coarse alignment on three target stars, punch those coordinates into the computer, and then take a series of finer alignments until the guidance platform was back in balance. At that point, the ship would once again know its precise orientation in the three-dimensional bowl of space. It was similar to the sighting work Lovell had done throughout the mission, but this time the computer was so confused that the alignments had to be re-established starting from scratch.

Lovell was able to complete the job in comparatively short order. When he was finished, he reflected that it was a good procedure for any Apollo commander to have in his back pocket. In the event that a future spacecraft ever suffered a catastrophic systems failure, the first thing he'd need to know would be how to get the ship pointed properly again.

☆ ☆ ☆

Avoidable errors like Lovell's were nothing compared to the unavoidable hazards that still awaited the spacecraft during reentry. The first critical step would come less than an hour before the ship made its initial contact with the atmosphere, when it would jettison the service half of the command and service module, the part with the engine and the long-term life-support systems. The spacecraft that had begun its journey at the top of a 363-foot stack of rocket would then be reduced to an eleven-foot-tall cone with a heat shield covering its blunt bottom and enough oxygen and power to keep the crew alive for just a few hours.

That capsule would collide with the atmosphere at nearly 25,000 miles per hour, and withstanding that high-velocity hammer blow would be only part of the challenge. The spacecraft would also have to thread its way into an entry corridor no shallower than 5.3 degrees from horizontal and no steeper than 7.4 degrees. That translated to a fifteen-mile-wide keyhole in the sky, which was an exceedingly small target if you were taking a bead on it from a quarter million miles away. On a far smaller scale, if the Earth were the size of a basketball and the moon were the size of a baseball, the two worlds would be positioned twenty-three feet apart and the fifteen-mile-wide reentry corridor would be no thicker than a piece of paper.

The wages of missing that target would be immediate. Come in too steep and the crew would be killed by the g-forces, assuming the spacecraft didn't get shaken apart by the aerodynamic violence first. Come in too shallow and Apollo 8 would skip off the skin of the atmosphere and rico-chet into the void forever. Even a successful reentry would require riding the fire to splashdown in a way no crew had ever done before, with the temperature on the heat shield climbing to 5,000 degrees Fahrenheit—or twice the melting point of steel—which was much more daunting than the 3,000 degrees of orbital reentry. The intense heat would cause the spacecraft to be surrounded by a cloud of ionized gas that would be impenetrable to radio signals, meaning that this life-or-death maneuver, like so many others on this mission, would take place during a blackout.

And there was still more to worry about. Assuming that the crew could hit the ballistic bull's-eye, the reentry would not be survivable if the spacecraft didn't do some complex maneuvering on the way down. The faint wisps of the upper atmosphere wouldn't present much resistance until the ship had descended to about 400,000 feet above the Earth, or about 75 miles. At this point, the spacecraft and crew would be pulling 2.5 g's, an easily survivable load. Eventually, the g's would climb to between 6.8 and 7.0, which would be tougher but still manageable. But after that, if the ship continued on the same trajectory, the g-forces would multiply lethally.

Instead, the command module would have to pull up, climb briefly back out of the atmosphere, allow the temperature of the heat shield to cool a bit and the g-forces to vanish, and then re-reenter at a shallower angle. The physics of the so-called skip reentry were not unlike those of a roller coaster: the first plunge is always the steepest, and each successive peak and valley is lower and shallower, as the gravitational energy that is accumulated during the slow, clanking climb to the top of the ride's first peak is steadily dissipated.

On the chalkboards and notepads where the skip reentry was first worked out, it all made clean, unarguable sense, except for one small thing: the Apollo spacecraft had no wings. Without wings you can't have lift, and without lift you can't climb. But there was an answer to that, too: design the command module so that its center of gravity was deliberately off-kilter.

Rather than placing that invisible pivot point at the center of the spacecraft cone, the engineers positioned it above the center. That design feature led to what was called a stable trim attitude, which naturally positioned the ship with its heat shield at a shallow downward angle. This is how the command module would be pointed as it made the first part of its plunge. When it came time to climb, the spacecraft, governed by the computer, would roll over, with the heat shield still facing forward but the astronauts now upside down in their couches. That would put the center of gravity below the center line, changing the angle of approach and causing the capsule to climb. When it was time to dive back down, the ship would roll once more.

The skip reentry would be a wild ride, and it would have been awfully nice to have test-flown the flight path from just a few thousand miles above Earth once or twice before. That was exactly the test Frank Borman, Jim Lovell, and Bill Anders had been scheduled to run when Apollo 8 was still called Apollo 9 and the crew was slated to bring a LEM and not go anywhere near the moon. But the ship never got its LEM and it went to the moon instead, and now the astronauts' test flight would be a real flight from a true lunar distance.

And that's why Frank Borman did not care a whit for talk of parades and world tours and relaxed vigilance in Houston.

☆ ☆ ☆

It was deep in the Pacific night when the spacecraft from the moon was making its final approach to Earth. If you'd lived on one of the islands in the middle of the ocean and stood outside at about 3:00 a.m. local time on December 27, you could have seen the ship coming, though you might have needed the help of a pair of seven-power binoculars. Apollo 8, through those lenses, would have been a faint point of light about a quarter of the way between the moon and the bright point that was Venus. You would have had to be patient, though: if you saw the target dot move a tiny bit over the course of fifteen minutes, you'd know you were looking at the spaceship, which by then would be just over sixty minutes from entering the Earth's atmosphere, moving at about 20,000 miles an hour and still accelerating.

Aboard the dot, the experience was very different and much less peaceful. The men in the machine couldn't feel the ship's motion, but the Earth, which had been as small as a coin just two days before, was growing fast, expanding well beyond the frames of the spacecraft's windows and once again becoming an enormous arc of horizon that was far too big to be seen whole in a single field. From the distance of the moon, Jim Lovell had marveled that he could extend his arm and hide the entire Earth behind his thumb. Now, as the planet resumed its proper scale, his thumb was once again a near nothing against the great mass of the world.

Before the crew could fly much closer to that growing bulk, they had a lot to do. As soon as reentry caused even a ghost of gravity to appear, any loose debris that had been floating around the cockpit would fall down to the base of the spacecraft, which meant onto the astronauts. This problem would only get worse as the g-forces grew, and you did not want to get hit in the head by a seven-g flashlight or bolt when you're trying to pilot a spacecraft through free fall.

Borman and Lovell did a fast cleanup of the vehicle. Anders, meanwhile, made sure the valves on the potable water tank and any coolant systems or evaporators were closed, since water raining down on the electronics would be even worse than junk raining down on the crew. Lovell watched Anders to ensure that the job got done and then called the ground to confirm.

"Bill just shut the potable inlet, Ken," he radioed to Mattingly.

"Okay, thank you."

"If I see any water floating around, I'll give you another call," Anders added.

The comment was meant to reassure, but it did no such thing. By now, any loose water that hadn't shown itself probably wouldn't appear until it came out of hiding during reentry, and then it would be too late.

The ground had other chores for the crew to perform, too, including making sure that the space suits, with their heavy fishbowl helmets, were securely stowed. If one good thing had come out of Wally Schirra's refusal to wear his helmet during Apollo 7's return to Earth, it was that the uneventful reentry allowed NASA to feel confident that the spacecraft's design was solid and the vehicle was unlikely to spring a depressurizing leak during the return. As a consequence, Mission Control gave Borman and his crew permission to reenter in the lighter, much more comfortable cloth jumpsuits they had worn throughout the flight.

Anders also had to adjust the environmental control system so that the cabin temperature fell to 62 degrees. That was too low for their thin garments, but the chill wouldn't last long. Even the best-insulated ship would experience some heat soak when its leading edge was reaching 5,000 degrees. And by the time the fiery reentry was over, the ship would have traded the deep freeze of space for the 85-degree weather of the near-equatorial South Pacific, at which point it might grow uncomfortably warm inside.

Though the skies over the splashdown area were due to be clear, the weather service was also forecasting four-foot waves. That prompted Mattingly to offer one further advisory.

"It has been recommended that since Marezine takes some time to take effect, you might want to consider taking one now," he radioed up.

Borman could not believe what he was hearing. Marezine—again. He had flown planes through chop and so had his crewmates, and this would be just one more bumpy ride. Wisely, the capcom had provided himself with some cover, mentioning in the same transmission that he had another set of figures on the entry PAD to read up to the crew. Borman decided that he would address only one of Mattingly's points, the preliminary advisory data.

"Okay, stand by. Let me get out the entry PAD," he answered, making no mention of the Marezine. Mattingly wouldn't mention it again, either.

Finally, the time came for the no-turning-back maneuver—the junking of the service module. Like so much else during this voyage, it would be an exercise in controlled violence, with explosive bolts blowing all links between the two parts of the ship. Immediately afterward, the ground would send up a command for the now-headless service module to fire its forward thrusters so that it could back away from the command module and then tumble to its death in the atmosphere.

"Houston, Apollo 8, confirm go for pyro arm," Lovell called down, asking for official clearance to arm the pyrotechnics that would effectively pull the pin on the jettison grenade.

"Apollo 8, you are go for pyro arm," Mattingly answered.

With that clearance from the ground, the astronauts returned to their seats and cinched their restraints as tight as possible. The loose settings used during the LOI and TEI burns would be nowhere near sufficient for the ride that lay ahead. After Borman belted himself in, he glanced over to Lovell and Anders to make sure they had done so as well. Then he looked at the clock on his instrument panel. The moment the service module was gone, he and his crew would be entering the period of peak risk. Within the hour, they would be safely in the water—or they would be casualties of space.

☆ ☆ ☆

On the ground in Houston, it was just after 8:00 a.m. Television coverage, which had returned to the air on all three networks, was intercutting between the voices of the anchors in their studios and the voice of Paul Haney, the Mission Control commentator currently on duty. Haney or one of his colleagues had been on the microphone for the entirety of the mission, but for much of that time, the newscasters had dominated the broadcasts. Today, the TV personalities stepped back to give more room to the voice of the space program.

"The route of flight, in case you're not looking at a map," Haney said in the flat, unemotive voice all of the NASA commentators used, "will be over northeast China, Peking, and then over Tokyo, and then we start a southeastern slant. The landing point is at one hundred and sixty-five degrees west, approximately eight north. That point, by the way, is just six hundred miles northwest of Christmas Island, which I'm sure has been noted." He was referring to the Kiritimati atoll, south of Hawaii, which sometimes goes by the name of the better known Christmas Island much farther west. The reference was a stretch, but one Haney was willing to make given the season and circumstances.

With just ninety seconds left until the service module would be jettisoned, Mattingly could be heard on the air alerting the crew that the spacecraft's primary evaporator had dried out. As with the evaporator problem that occurred during the first broadcast from lunar orbit, this issue would ordinarily require immediate attention. But since both the primary and the backup evaporators were located in the service module, they had a mere minute and a half to live. Mission rules, however, were not tolerant of excuses, so the command to switch to the backup system was sent up to the ship. Anders radioed down a bemused "Roger," and it was left to Haney to offer the viewers some candid perspective on the exchange.

"Crew's been advised that their primary evaporator has dried out," he said, "a fact that I'm sure that they couldn't care less about."

In his center seat, Lovell tapped the commands that would initiate the separation sequence into the computer. After thinking for a moment, the computer processed what it had been told and flashed back the 99:20 "go or no-go" code.

"Go to proceed," Lovell called out.

"Go to proceed," Borman agreed. He placed his hand on his thruster controller in the event the maneuver went awry and the ship flew off course.

Lovell pressed the PROCEED key. As the bolts exploded, a dull thump accompanied by a jolt shook the astronauts. The command module popped free, and the service module—drifting somewhere invisibly behind—fired the proper burst through its forward-facing thrusters and backed safely away.

Borman eyeballed his attitude indicators and relaxed his hand. His ship was stable.

"*That* was a kick," he said. Though he was fully aware of the power of the pyros, he was surprised that he had felt so much of the explosion's force.

Mission Control could see from its telemetry that the jettison had occurred, and Anders and Lovell had felt it as powerfully as Borman had. As a consequence, no one thought to say it aloud.

On Earth, the newscasters got jumpy.

"Separation should be taking place just about now," Cronkite said. "We're waiting for confirmation of that from Paul Haney." But Haney wasn't saying, and after more than a minute had elapsed, Cronkite fretted. "We're *still* waiting for confirmation of the separation, which should have taken place at thirteen seconds after 10:22 eastern time," he said. "To make us all happier, we'd like to have it."

At last, Haney either realized his oversight or was told by someone monitoring the networks to make the anchorman happy. "The flight director has confirmed separation," he said.

"So apparently all things are going well with the flight of Apollo 8," said Cronkite, obviously relieved.

Reentry was now an inevitability less than twelve minutes away. But the success of that imminent collision between spacecraft and air required one more navigational sighting.

Precisely six minutes before reentry began, the moon—remote once more—would rise over the horizon of the Earth, the last time the crew would see it through the void of space. If it showed itself at the expected moment, it would mean Apollo 8's trajectory was true. If not, Lovell and Borman would have some fast and complicated navigating and flying to do to put their course right before time ran out.

Two minutes before the lunar sighting was due to occur, Anders, following the flight plan, called out, "Horizon check."

Lovell, who had been watching Borman from his adjacent seat, responded: "He's doing that now."

Anders, confident that the moon sighting would go as planned, began reading off the next set of commands. "Pitch needle error goes toward zero," he said.

Borman, keeping his eyes on the horizon outside, answered only, "Okay."

"Don't forget manual attitude to three, rate command," continued Anders.

"Yes, okay, but tell me that later," Borman replied.

"Yes, right. Just don't forget it."

"Just tell me later," Borman said. "Okay?"

The commander looked away from the horizon and scowled at his instruments. The ship was fighting the proper line of descent, straining up and away from the route it needed to travel. Suddenly the scheduled moonrise was a secondary concern.

Borman fired a burst from his thrusters.

"See where this baby wants to fly? The pitch is way up," he muttered. Then he fired the thrusters again and began bringing the spacecraft to heel. "Keep checking my yaw for me there, will you?" he asked Lovell.

"I will," Lovell answered. "You're a little bit left."

Borman brought the ship back to midline.

"It looks like you're slightly rolled," Anders said.

"I don't care about the rolls," Borman said—caring indeed, but still monitoring yaw. A moment later, he corrected the roll too, and the ship finally seemed to stabilize.

Anders checked the indicators. "We've got a good horizon," he confirmed.

Borman, pleased, looked through his window and smiled. "And look who's coming there, would you?"

"Yes!" Anders exclaimed, glancing ahead.

"You see it?"

"Yes."

"Just like they promised."

"What?" asked Lovell, who had been minding his instruments.

"The moon," Borman and Anders said in unison.

Lovell looked up and saw it, too—a tiny world that, like the Earth a few days ago, could fit behind his thumb.

"At six minutes before," Borman said, glancing at the flight plan, "just like it says."

Haney, reading the telemetry that indicated the ship's stable attitude, now described for the television audience the coming sequence of events.

"The four-hundred-thousand-foot point is where they will begin to encounter some little bit of atmosphere," he said. "The blackout should begin some twenty-five seconds later. The max g-force felt by the crew will be six-point-eight g's. A second g spike of about four-point-two will be noted about four, five minutes later. The total blackout we're predicting this morning is on the order of three minutes. But since we have very little experience reentering at these velocities, we must caution you that those are only estimates."

It was all so informal—the list of events that would happen in a predictable order, the g-forces that would be merely *noted* by the crew. But for Haney to say that NASA had "very little experience reentering at these velocities" was like saying that until Apollo 8, human beings had very little experience flying to the moon. In fact, they had had none.

Inside the ship, the astronauts kept their tone equally nonchalant. Lovell looked through his window and noticed the way the onion skin of the atmosphere now looked so much thicker—thick enough that the sun shining through it broke into a spectrum of shades from the black of space to deep blue, then red, orange, and finally bright yellow. Just three Christmases before, he and Borman had spent fourteen days with that rainbow ribbon outside their window; Anders, who had been extremely busy during Apollo 8's brief stay in Earth orbit six days ago, had likely not seen it at all.

"I got the old . . ." Lovell began, waving his hand toward the window.

"What is that?" Anders asked.

"Good old airglow is what it is," Borman said.

Anders glanced at it; unimpressed, he looked back down at his flight plan. "I'll look at the airglow next time," he said.

Lovell pressed: "You've never seen the airglow. Take a look at it."

"You can't get your pin without seeing the airglow," Borman teased, echoing the stewardesses on airplane flights who would bribe restless children into good behavior with the promise of the gift of souvenir pilot's wings.

"That's right," Lovell said.

"I see it, I see it!" Anders laughed, making a show of gaping out the window. Then he affected the nervous look of the rookie pilot. "Let's see, is this where I'm supposed to ask, 'How many g's, Lovell?'"

Anders could joke, but the fact was, the needle on the g-meter had indeed begun to twitch. Meanwhile, the airglow was growing bright enough to reflect off the windows. The reflection quickly grew brighter and redder, the first sign that the ship was encountering air resistance.

At that moment, NASA radioed up a reminder to the crew to turn on the radar transponder that the recovery ships would need to track the spacecraft as it fell. But Mission Control would have to trust that the astronauts had heard the command and obeyed it, because the instant the words were uttered, the communications link was cut.

"And we have lost signal," Haney announced.

☆ ☆ ☆

Lovell didn't need a mission commentator to tell him what the hiss in his headset already had. The crew was alone once more. He turned to Anders.

"You've got the checklist again, Bill?" he asked.

"You've got it?" Borman repeated.

"Yes," Anders assured them, holding up the flight plan.

"I'll tell you when the g's start going," Lovell said, the matter no longer a subject of humor.

"This is going to be a real ride; hang on," Borman said, then turned to Lovell. "You got point-zero-five g?" That reading—just .05 of a single g—would be the first data reading indicating that reentry had officially begun.

"I've got point-zero-two," Lovell said.

A stray washer, missed in the cleanup but now with a tiny bit of weight to it, drifted into view, falling in slow motion.

"There goes a washer," Borman said. "Can you grab it?"

Before anyone could get it, the little bit of junk floated off again.

Lovell's eyes remained on his instruments, watching both his g-meter and the mission clock. He knew precisely how the joint forces of gravity and time were supposed to play out. "Stand by—thirty-eight, thirty-nine, forty, forty-one . . ." His voice trailed off. And then: "Point-zero-five!"

"Point-zero-five," Borman confirmed.

"Okay, we got it!" Anders said.

"Hang on!" Borman called.

"They're building up," Lovell said.

"Call out the g's," Borman commanded.

"We're at one g!" Lovell called.

The astronauts stayed silent for just twenty seconds as the g-forces rapidly multiplied. Lovell watched the needle climb to two, then three, then four and beyond.

"Five," he said, straining to get out the syllable as a force quintuple Earth's nominal gravity pressed down on his chest. Then yet another rock was piled on.

"Six," he said through his teeth.

The red glow outside the spacecraft brightened to the orange of a roaring fire, then to deep yellow, then brilliant yellow, then a pure, nearly blinding white. The astronauts squinted against it; to Borman, it was like being on the inside of a fluorescent bulb. As the white light, which had no other degree of brightness left to it, held steady, the g's, which knew no such limits, climbed to a peak of 6.84.

Then, at last, the roller coaster rose according to the plan and the g's started backing off.

"Four," Lovell said, relieved at that minimal easing of the pressure.

"Quite a ride, huh?" Anders said.

A moment later, Lovell, breathing more easily still, provided another report: "We're below two g's."

"Nice job there, gang," Borman said.

The break would not last. The final dive was yet to come, when the g's would climb back over four. The astronauts were still more than 175,000 feet—or 33 miles—above the Earth and free-falling through the sky. Not until 24,000 feet would the two thirteen-foot-diameter drogue parachutes be deployed, jerking the spacecraft to a slower but still lethal 200 miles per hour. Only at 10,000 feet would the three main chutes, each 83.5 feet across, open up, braking the spacecraft to a tolerable land- ing speed.

Even then, "tolerable" was a relative notion. The spacecraft would hit the water at just over 21 miles per hour, which seems like nothing when you've been flying a ship that just minutes before was moving at one thousand times that speed. But 21 miles per hour is also thirty-one feet per second, a speed that on impact makes water feel like a solid and shakes you with a force that rattles your teeth. The arrangement of the parachutes was designed to mitigate that a bit, since they were attached in a way that caused the spacecraft to hang from them somewhat crookedly, so that the command module didn't slam into the water on its wide, flat bottom but sliced into it with the leading edge first. Further, the astro- nauts' couches were mounted on crushable aluminum struts that were meant to collapse on impact and thus absorb some of the blow.

But all that was for when Apollo 8 was a lot closer to the Earth and moving a lot slower than it was right now.

☆ ☆ ☆

Far below the spacecraft, rescue helicopters had already begun scram- bling from the deck of the USS *Yorktown* in the Pacific, converging on the site where the command module was likely to splash down. Once the rescue teams arrived, there would be nothing for them to do but hover and maintain their positions.

For now, Apollo 8 still hadn't emerged from blackout. The ship was at least a minute away from that milestone, and on the ground Haney did what he could do to manage expectations.

"Our curves put the spacecraft down about thirty-five to thirty-six miles above the Earth," he said.

In the background, Haney could hear Mattingly make a preliminary call to the ship, and if he could hear it, the viewers heard it as well. They needed to know not to expect a reply. "Ken Mattingly just put in a call and frankly labeled it a radio check," Haney said. "He's gotten no response as yet."

Cronkite, decidedly not managing expectations, broke in and spoke more bluntly. "If the blackout ends as planned," he said, "it should be over in about ten or eleven seconds from now."

Ten and then eleven seconds ticked off, but there was still no word.

"Apollo 8, Houston, through Huntsville," Mattingly said, trying to reach the ship through the communications network at NASA's Huntsville, Alabama, facility.

Again he got no response.

"And Ken puts a second call in to the crew," Haney said. "About three and a half minutes since we went into the blackout."

Moments later, word flashed through Mission Control that Huntsville had just picked up a first radar signal from Apollo 8. It wasn't much, but it meant the ship was alive.

Haney went wide with the news. "And Huntsville says that they have acquired an S-band signal," he announced.

Cronkite jumped on the report. "They do have a sig—"

Haney cut him off. "They immediately called back and said no contact," he said firmly. "They negate that announcement."

Cronkite groaned audibly. "I really thought we had them," he said.

The blackout now stretched to four minutes. In the Borman, Lovell, and Anders homes, the only sounds came from the television sets and the squawk boxes. The children—eleven of them among the three families—were all awake, and those who were old enough to understand what was happening were watching the coverage. For these final moments of the mission, all three mothers were sitting in their living rooms, their children close by.

The blackout stretched to four and a half minutes and then closed on five.

"It's now just two minutes past the time we should have heard from the spacecraft," Cronkite said somberly.

"Houston, Apollo 8, through Huntsville," Mattingly called again.

He allowed almost another minute to pass.

"Apollo 8, Apollo 8, this is Houston," he tried once more.

Fifteen more seconds of silence elapsed.

Then, at last, through the loud wash of air-to-ground static, Jim Lovell's voice—broken but discernible—filled the Mission Control headsets and the living rooms around the planet.

"Houston, Apollo 8, over," he said.

"And . . ." Haney began, his voice choking before he took a second breath, "there's Jim Lovell!"

"Ha-ha!" Cronkite exclaimed.

"Go ahead, Apollo 8, read you broken and loud," said Mattingly.

"Roger," Borman shouted back through the crackle and roar of the plasma cloud only now dissipating from around the ship. "This is a real fireball! It's looking good!"

"He says we're looking good!" Haney said.

"It's almost all over but the shouting, men," Borman said to Lovell and Anders.

The altimeter on the instrument panel showed that the ship was fast approaching 24,000 feet, meaning the drogue chutes were about to be deployed.

"Make sure your heels are locked," Anders called to Borman and Lovell, reminding them of their training.

Borman looked out his forward window. "There she comes!" he said, as the conical nose of the capsule blew away and the two red-and-white drogue chutes reefed and bloomed.

The astronauts jerked back in their couches, but the tenuous communications link had broken off again, and the crew did not even bother to call the ground to confirm the chute deployment.

Anders watched as the altimeter, dropping more slowly now, continued toward 10,000 feet.

"Should be approaching ten K," he said a few moments later. "Stand by with the mains in one second."

That second elapsed, and at just over the promised altitude of 10,000 feet, the main chutes burst free. The men jerked once more, and the ship's descent slowed dramatically.

The communications loop opened up again and another voice—an entirely new one—filled their headsets.

"This is Air Boss 1," someone announced, using the call signal of a rescue helicopter. The unmistakable sound of rotor blades chopped in the background. "You're sounding very good, very good. You have been reported on radar as southwest of the ship, about twenty-five miles."

"Roger," Lovell answered.

The spacecraft fell through 8,000 and 6,000 and then just 5,000 feet. After a journey of half a million miles, the ship was now less than a mile above the water.

"The spacecraft is down to one thousand," the pilot of the helicopter called to the *Yorktown*.

"Brace yourselves," Borman yelled to Lovell and Anders.

"Welcome home, gentlemen," the pilot said, a few seconds prematurely. "We'll have you aboard in no time."

"Stand by," Borman told his crew. "Stand by for Earth landing."

A moment later, the three astronauts felt the hard hand of the Earth's surface hit them at their backs as their spacecraft half-sliced, half-slammed into the rolling waters of the Pacific Ocean. The struts beneath their seats collapsed as they had been designed to do, but the jolt was still violent.

The crew barely noticed. Borman pumped his fist, Lovell and Anders let out a whoop, and all three men looked at one another and grinned.

"*Yorktown*, Recovery 3," a helicopter pilot called. "At this time the spacecraft is in the water."

Walter Cronkite, his voice filled with relief and jubilation, made it official. "The spacecraft Apollo 8 is back!"

☆ ☆ ☆

And so it was. Getting the spacecraft into the ocean was, however, not the same as getting the astronauts onto the carrier. At that moment, about four miles separated Apollo 8 from the ship. And distance wasn't the only obstacle; time was, too. Splashdown had occurred at 4:51:50 a.m. Hawaii-Aleutian time, or 10:51 a.m. on the East Coast.

It would not be difficult to find Apollo 8 in the predawn darkness, thanks to both the flashing white light on the nose of the ship and the radar beacon that had been pinging throughout reentry. But Pacific sharks prowl during the early morning hours, and nobody wanted to take the chance that either the moon men or the frogmen would attract their deadly attention. So the astronauts would have to bob and drift in their airtight pod until first dawn, which was still at least half an hour away. Once the skies brightened, it would take another hour for the frogmen to hit the water and attach a flotation collar to the spacecraft, which would make it possible for the astronauts to leave their ship without falling into the sea. Only then could the hatch be opened and the crew extracted, at which point the men would be lifted one by one up to the helicopter that would fly them back to the *Yorktown*.

The wait for their rescuers would not be pleasant. NASA's meteorologists had been right when they'd forecast waves in the recovery area, but they had been wrong in predicting that the waves would top out at about four feet. In fact, they were closer to six. And though it had been a good idea to turn the cabin temperature down to 62 degrees before reentry, that interior chill had been replaced by a stultifying heat, first from the fiery descent through the atmosphere and then by the warm temperatures of the near-equatorial Pacific. The combination of the two— hot, close air and the heaving of the capsule—did not do good things to a pilot's insides.

Making the crew's circumstances worse, the motion of the waves soon caused the spacecraft to flip from the position NASA called "stable one" to the position called "stable two"—what anyone else would call right side up to upside down. That left the astronauts hanging in their straps in the now unfamiliar one-g environment, looking down at their instrument panel. Borman could activate three flotation balloons packed in the nose of the spacecraft near where the parachutes had been, and he promptly did so. But it would take a while for the balloons to turn the spacecraft back over. Both Mission Control and the recovery helicopters noticed relatively little chatter coming from the spacecraft, a silence characteristic of anyone battling a rising bubble of nausea.

"Get us out of here," Anders finally radioed, only partly joking. "I'm not the sailor on this boat."

Lovell, the Navy man flying with two Air Force men, managed to keep any seasickness in check. So did Anders. Borman, who had begun his mission to the moon fighting a losing battle against his rebellious stomach, lost again. If the Apollo 8 spacecraft was destined for a museum—which it surely was—the conservators would have a little cleanup work to do first.

Eventually, the sky did brighten, the choppers closed in, and the frogmen leapt into the water. They waved at the crew through windows that less than seventy-two hours before had been filled with the bright, bleak landscape of the moon; now, seeing the faces of smiling strangers, the astronauts waved back. Once the flotation collar was secured, one of the frogmen signaled the all-clear by banging on the hatch. Lovell, in the center position—the position occupied by Ed White in a different ship almost two long years ago—opened the little door with ease.

The fresh smell and warm breeze of the Pacific flowed into the capsule, replacing the stuffy air the crew had been breathing for nearly a week.

"Welcome home, men," the lead frogman said.

The astronauts were helped out of the spacecraft, onto the docking collar, and into a life raft—Lovell first, then Borman, then Anders. As a helicopter hovered noisily overhead, a rescue basket lifted them to safety. The chopper's crew saluted each man as he climbed aboard.

"Congratulations, sir," said Donald Jones, the pilot, to Borman. "It was a wonderful reentry."

Borman demurred. "It was all automatic," he said. "We had nothing to do with it."

The commander of the first mission to the moon was pretty sure that he would be congratulated and applauded a lot in the coming weeks, but he had already decided that he would take credit only for the accomplishments that were actually his. Besides, he had something else on his mind.

"Did anyone bring the shaver?" he asked.

"Right here, sir," one of the young crew members said, handing Borman an electric razor.

Borman had taken a lot of ribbing over the patchy blond beard he had grown during his two weeks aboard Gemini 7; during the same mission, Lovell had practically grown a full beard. Now the Apollo 8 commander flicked on the razor and cleaned up the stubble that had sprouted over the past six days. No one would have anything to tease him about this time.

When the helicopter at last landed on the *Yorktown*'s deck, a red carpet was waiting. The door opened and the astronauts emerged, grinning and waving as the carrier's crew cheered. In Houston, the mission controllers had scrupulously abided by the rule that no final celebration would be permitted until that moment. When they saw the astronauts appear, they, too, shouted and hugged and shook hands and lit cigars.

The astronauts walked between the ranks of cheering sailors, waving and calling out thank-yous. At the end of the carpet, they were greeted by Captain John Field, the commander of the ship, who presented them with USS *Yorktown* baseball caps, which they donned as demanded by both tradition and a genuine gratitude for the efforts that had been made on their behalf. The astronauts shook hands with Field, who then motioned Borman to a standing microphone that had been placed there for him.

Though Borman had not prepared a speech, he welcomed the chance to show his appreciation to the carrier's crew.

"It seems that Jim and I always fly in December, doesn't it?" he said, to appreciative laughter from the ship's crew. "That time we got back before Christmas. This time we didn't, and we want to apologize for keeping you out here over the holidays."

After a few more remarks and thank-yous, the astronauts vanished belowdecks to the sick bay for their postflight medical exams. There would be a call to take from President Johnson and another from Vice President Humphrey, and already congratulatory statements and telegrams were streaming in for the astronauts from world leaders— U Thant at the U.N. and Queen Elizabeth at Buckingham Palace and Prime Minister Wilson at 10 Downing Street, as well as the presidents of France, Italy, and many other countries. The Kremlin sent its congratulations; in fact, the White House had used the Washington-Moscow hotline—installed mostly as a direct source of communication during times of military escalation—to keep the Soviets apprised of the status of the mission.

"Accept, Mr. President, our congratulations on the successful completion of the flight of the Apollo 8 spacecraft around the moon," said Soviet president Nikolai Podgorny in his formal statement to Lyndon Johnson. The mission, Podgorny noted, "is a new accomplishment in mastering the outer space by man." The telegram implicitly acknowledged that the United States had just won a dramatic battle in the Cold War, and the official spokesman for the opponent was—grudgingly perhaps, but graciously—conceding that fact. In a contest so often marked by ambiguous outcomes, Apollo 8 had scored a clear victory.

Belowdecks, freshly showered, the astronauts had changed into clean white flight suits, combed their hair, and donned their *Yorktown* caps, and soon they were climbing the metal ladder back to the deck. By now, their spacecraft had been hauled aboard, and the portion of the deck that held the little capsule had been cordoned off.

The astronauts approached their spacecraft and examined it. By any measure, the ship was spent. Its flanks had been discolored by the fires of the reentry, and its heat shield was half incinerated. Its tapered nose had been jettisoned to allow the parachutes and flotation balloons to deploy, giving the front of the capsule an incomplete, even broken look.

The hatch was partly open, and the men looked inside. It was a cluttered mess. There would be a few souvenirs to collect—the flight plans, a flashlight, the little unopened bottles of brandy, perhaps. But the command module was now retired, an artifact of history.

Jim Lovell and Bill Anders regarded their spacecraft with an acute feeling of unfinished business. If given the chance, they would take another ship—tomorrow wouldn't be too soon—and fly the whole mission again, only this time they would land on the moon they had so recently orbited.

Frank Borman felt nothing of the kind. He had been assigned a mission, he had flown the mission, and he had come home from the mission. He gave the scarred side of the command module an affectionate slap, then turned around and walked away. He did not look back.

EPILOGUE

January 9, 1969

NOBODY IN WASHINGTON USED THE TERM "JOINT SESSION" TO DESCRIBE the appearance by the Apollo 8 astronauts before both houses of Congress, even if most of the newspapers did. A joint session of Congress, as Capitol Hill's parliamentary rules make clear, happens only after both the Senate and the House of Representatives pass a momentous resolution—a declaration of war, say—and then gather to hear an address, usually by the president. Instead, the Apollo 8 astronauts were invited to appear before a joint *meeting* of Congress, a formal event that required no joint resolution but was simply a gathering in the chamber to hear a speech by an exceedingly distinguished person or persons. Semantics aside, however, the scene was the same.

On that day, Frank Borman, Jim Lovell, and Bill Anders stood at the Speaker's rostrum at the front of the House Chamber. Attending the meeting were 100 senators and 435 representatives, assuming all of them showed up—and from the look of it, no one had called in sick. Vice President Hubert Humphrey and Speaker of the House John McCormack sat behind the astronauts, in the same kingly chairs they used when the president was speaking. The members of President Johnson's outgoing cabinet had taken their usual spots near the front of the chamber, as had the Joint Chiefs of Staff and the robed members of the Supreme Court.

The three astronauts looked fit and happy as they stood and waved, and if you somehow didn't recognize their faces after all the recent press attention, you would never know that they were less than two weeks removed from having been to the moon. If you did recognize them, there was something discordant about seeing them in this place, wearing these clothes. Men who had almost never been photographed in anything other than pressure suits or in-flight jumpsuits had swapped their work clothes for conservative business suits. The effect was like seeing a World Series star out of uniform, dressed for a formal dinner or an awards ceremony. No matter how hard he tried, he just didn't look right.

But none of that mattered, because the politicians couldn't get enough of the spacemen. They came to their feet and cheered the minute the Apollo 8 crew appeared at the back of the chamber; they applauded wildly as the three men walked down the aisle, smiling and shaking hands just like a president does. The legislators listened closely as each man spoke in turn about the journey he had just completed and about America's preeminence in space. They listened especially closely when Borman spoke. He kept his remarks to just twelve minutes, but he made each minute count. No one missed his meaning when he thanked the legislators for the $24 billion they had approved over the years to fund the Apollo program, and everyone knew that a significant number of legislators in the chamber that day had voted against that funding.

Borman also spoke of his belief that there would soon be small colonies of scientists living and working on the moon, much the way they did in Antarctica. "I'm convinced it is no longer whether we'll do these things," he said, "it's a question of how long it will take and how much we'll spend. Exploration is really the essence of the human spirit, and I hope we will never forget that."

His speech was simultaneously tactical and inspirational, just the kind of remarks you might hear from a politician—and, indeed, several people were already sizing up the forty-year-old space hero as a prize political racehorse, a man who could win practically any office he fancied. And like any good politician, Borman used the occasion to make a

joke at his own expense. Commenting on the crew's Christmas Eve
reading from Genesis, he looked down at the justices of the Supreme
Court—a court barely seven years removed from having ruled prayer in
the classroom unconstitutional—and said, "But now that I see the gentlemen in the front row, I'm not sure we should have read from the Bible
at all." He went after the laugh and got it: his audience roared.

☆ ☆ ☆

The country's celebration of the astronauts' achievement had begun
almost the moment they returned to Earth, and it would continue for a
good while. Shortly before speaking to Congress, the astronauts had met
President Johnson in the East Room of the White House and been
awarded NASA's Distinguished Service Medal. Then they had been
driven down Pennsylvania Avenue in a motorcade to the Capitol while
a great crowd of people waved and cheered. The astronauts' families,
who were invited to attend all of the events, rode in trailing cars in the
motorcade.

Marveling at the throngs of people, seven-year-old Susan Lovell
turned to her mother. "You'd think Dad was a hero or something," she
said.

"Well, Susan," Marilyn answered, "I think he is."

All three of the heroes got much the same treatment in New York
the next day. There were more medals and more dignitaries—Governor
Nelson Rockefeller and Mayor John Lindsay and United Nations
Secretary-General U Thant—and a formal dinner at the Waldorf-Astoria.
There was a ticker-tape parade, with all of the street signs along the route
changed temporarily to Apollo Way. As their procession crept along
lower Broadway in a 28-degree chill, two hundred tons of confetti fluttered down on the astronauts.

At NASA itself, the larger work of the Apollo program soon
resumed—but not before a brief spell of hard-earned, celebratory madness. *Time* magazine reporters, who had telexed almost thirty reporting
files back to New York during the last three days of the Apollo 8 mission
alone, devoted an entire final file to the parties. James Schefter, a one-
time reporter for the *Houston Chronicle* who had covered the Gemini

program and thus knew a thing or two about the ways Houston played, was unrestrained when reporting what he saw at the splashdown party thrown in a hotel ballroom by TRW Systems.

"They were happy. They were uninhibited. They were grandly, gloriously drunk," he wrote. "They were earnestly seducing secretaries in darkened corners, smooching unabashedly in the brightly lit lobby, lifting quarts of booze from the press party next door. Reporters and public relations men retrieved some of the missing liquor, put a guard on the rest, and swung into a fest of their own."

The technicians and mission controllers, funding their own party on government pay, rented out a more modest two-story restaurant a short distance away on NASA Road. Leaving their cars parked up and down the four-lane stretch of highway, they swarmed inside. Those who couldn't find even one of those barely legal parking spaces simply abandoned their cars on the median strip; the Houston police, for whom it was still just another work night, ticketed them with abandon.

One final bit of Schefter's reporting told a different story, however, one that was more in keeping with the way the world saw the people of NASA. After finishing his descriptions of the parties, Schefter wrote, "Over in Mission Control, a team of flight directors missed it all. Reporting for duty at 3:30 p.m. [on the same day as splashdown], they went into a ten-and-one-half-hour simulation for Apollo 9. And the cycle began anew."

So it did. After the success of Apollo 8, that cycle spun fast. On March 3, 1969—just over nine weeks after Borman, Lovell, and Anders splashed down—Jim McDivitt and his Apollo 9 crew went into Earth orbit and spent ten days flight-testing the lunar module. They fired its engines and checked its guidance; they ran the LEM's computer through all of its commands. McDivitt and Rusty Schweickart flew free in the untried machine and then docked with the command module, piloted by Dave Scott. At long last, the spindly ship that had given its engineers such fits was declared fit for service.

Apollo 10, commanded by Tom Stafford, followed in May. The NASA bosses—bold enough to fly Apollo 8 to the moon but still chastened by the hubris that had led to the fire—decided that one more dress rehearsal was needed before attempting a lunar landing. Stafford and his crew would thus pilot another lunar orbital mission, this time with a LEM that they would fly as low as 47,400 feet above the surface—or 8.9 miles—much closer than the Apollo 8's sixty-nine-mile orbit. So the first-man-on-the-moon profiles of Stafford were shelved, to be replaced by stories about Apollo 11 commander Neil Armstrong—and those were the stories that ultimately ran.

Initially, Armstrong's assignment had not been a sure thing. NASA, reflexively risk averse, was always looking for ways to improve the reliability of the hardware and the skill of the crews. After Apollo 8 returned to Earth, Deke Slayton and a few others in the flight planning office began talking quietly about the possibility of giving Borman, Lovell, and Anders a few weeks off to rest, then putting them right back into training and sending them off to fly Apollo 11. The crew was uniquely qualified for lunar flight, after all, and they would have at least seven months to prepare for their new assignment. And although the LEM would introduce an entirely new variable into the mission, no one knew its workings better than Anders.

Before the offer was even made to the Apollo 8 crew, however, Slayton and others thought better of it. Armstrong, Collins, and Aldrin had been training hard for the mission, and it would show a lack of faith in the men—to say nothing of the training program itself—to switch the assignments now.

Borman got wind of the rumors that he might be offered Apollo 11 and was relieved that they came to nothing. If other astronauts wanted to go to the moon to plant a flag, let them. His Cold War was over. Even if he had wanted to fly again, he knew it wouldn't be fair. He had already put Susan and their two sons through the wringer of Apollo 8, and he would consider it a shirking of spousal and parental responsibility to vanish back into training for more than half a year and then subject them to an even harder, longer mission.

When Lovell heard that the crew assignments would not change, he was as delighted as Borman. Had the crew been assigned to Apollo 11, he would have occupied the center seat again, meaning that he would once again orbit the moon but not land on it. That held almost no appeal. He would much prefer to pass up an immediate return, put his

name back into the flight rotation, and take his chances on commanding a later mission.

Anders, however, was not the least bit pleased when he heard that
he wouldn't be flying on Apollo 11. He had lost his chance to fly a LEM
once, and now, with a trip scheduled for just seven months later, he had
had a shot at a new one. But once again it had been snatched away, and
unlike Lovell, Anders suspected that this meant he would never get a
chance to land on the moon. Lovell had three flights on his résumé, but
Anders was still a relatively junior member of the astronaut corps. If he
ever got a second shot at the moon, NASA's byzantine rules for making
crew assignments all but assured him a center seat, and the competition
for mission assignments completely ruled out a third shot. Boxed in,
Anders accepted an appointment as executive secretary of the National
Aeronautics and Space Council, working directly with President Nixon
and Vice President Spiro Agnew to formulate space policy. It wasn't flying,
but if he couldn't walk on the moon, he would do his best to shape the
future of American space travel.

With Apollo 8 behind him, Lovell moved back into NASA's usual
three-and-three crew-rotation cycle. As a result, he was named to the
backup crew for Apollo 11, though this time he would be assigned to the
left-hand seat instead of the center. If anything happened to Neil Armstrong before liftoff, Lovell would become the first man on the moon.
After that, he would command Apollo 14. Waiting six missions for his next
real shot seemed like a stretch, but NASA was launching Apollos pretty
quickly, and Lovell might get to go back to the moon in as little as eighteen
months.

As it turned out, he got his chance even earlier. Alan Shepard had been
named commander of Apollo 13, but an inner ear problem had kept
him grounded since his brief suborbital flight in 1961. Recently he'd had
the condition surgically corrected, and he had been returned to flight
status. After so long a layoff, however, he needed more time to train than
he and NASA had originally believed he would. Consequently, Lovell
was asked if he would mind swapping missions, taking his crew up on
Apollo 13 and letting Shepard and his team have 14. Lovell didn't mind
a bit, feeling that if you had the chance, it was always better to fly sooner.
Before long, Commander Lovell—the man who had orbited the moon

in a spacecraft that had done everything right—would learn what happens when a ship does everything wrong.

☆ ☆ ☆

Borman made good on his decision not to fly again. He served, for a time, as second-in-command in the astronaut office under Slayton, but the position left him restless. With little left to accomplish at NASA, he preferred to try working in the private sector, something that, as a career military man, he had never done. Unsurprisingly, he had no shortage of suitors.

Ross Perot, the wealthy Dallas-based founder of the computer company Electronic Data Systems, had just taken the operation public and seen his stock price jump tenfold in a few days. With plenty of money in hand and a reputation for knowing how to make more, he contacted Borman almost immediately after the astronaut returned from the moon and offered him a position producing a television show in which on-camera personalities would debate possible answers to national problems. Borman didn't have much interest in TV, but the civic-minded nature of the new show appealed to him, and he promised Perot that he would think about it after he returned from the Apollo 8 world tour.

While Borman was away, Perot decided to sweeten the pot. Shortly after the astronaut returned from the tour, he got a call from his broker, who came straight to the point. "Congratulations," he said. "You're a millionaire."

Borman asked what he meant, and the broker explained that in recent weeks Perot had invested $1 million in his name in the stock market and that, like most investments Perot made, this one was doing fine. But the idea that such quick, unearned money might suddenly be his made Borman uneasy—and it made Susan even more so.

"Give it back," she said. "You can't take it or he'll own you forever." Borman agreed, called Perot, and said thank you very much, but he was pursuing other opportunities.

Barry Goldwater, the Arizona senator and the Republican Party's 1964 presidential nominee, came calling, too, urging Borman to run for the state's other Senate seat. Borman's name and Goldwater's considerable political throw weight would probably make the race easy. But though

Borman was flattered, the glad-handing and vote hunting that constituted so much of a politician's life held no interest for him. And to a man accustomed to tight chains of command and orders that were carried out immediately, the Senate's exceedingly deliberate pace would prove maddening.

Finally, Borman was introduced to Floyd Hall, a former Army Air Corps flier and commercial pilot for TWA who had become the chief executive of Eastern Air Lines a few years earlier. Hall liked Borman and offered him a position at Eastern in the flight operations division. The job paid barely a third of the salary of some of the other opportunities Borman had been considering, but it involved airplanes—machines he knew a thing or two about—and an industry he could respect.

Borman accepted, and just over six years later, in 1976, he became Eastern's CEO. He led the company to its three most profitable years in history. In 1983, at a time when deregulation was disrupting the entire industry, Eastern's shareholders and board voted to sell the company to a larger airline conglomerate. Borman stayed on as CEO for three more years before retiring.

All three Apollo 8 astronauts earned substantial material wealth in the private sector, which they never could have done in the military or at NASA. Like Borman, Anders eventually occupied a CEO's chair, in his case for General Dynamics Corporation. Lovell thrived in the telephone communications industry, working for a number of companies in executive positions and riding the very profitable transition from landlines to wireless communications. But as is almost always the case with young icons who live long lives, all three men would remain far better known for their less material achievements.

★ ★ ★

The Apollo 8 crew returned to Earth on the very day *Time* magazine was finishing its annual Man of the Year issue, and whoever the editors had chosen—the betting was it would be Richard Nixon, the new president-elect—that person was summarily bumped. Instead, Borman, Lovell, and Anders appeared on the cover just a few days later, one of the few times in the magazine's history that the singular Man of the Year honorific was replaced by the plural.

Over time, Bill Anders's *Earthrise* photo became the defining image he'd sensed it might be, reproduced hundreds of millions of times on postage stamps, wall posters, T-shirts, coffee mugs, and more. Both *Time* and *Life* magazines ranked it among the hundred most influential photographs in history, and the image would widely be credited with animating the environmental movement, which was just beginning to gather momentum in 1968 and became a global force within the year.

The Apollo 8 crew would, in later years, become known around NASA as the Original Wives Crew. In a profession in which divorce was common, all three men still remain married to the same women they wed when they were young, and all three couples have broods of grandchildren and—in the Bormans' and the Lovells' cases—great-grandchildren to show for it.

Frank and Susan Borman eventually resettled in Billings, Montana, to be near their son Fred, who runs a ranch in Big Horn County, and to take advantage of the open-sky country, where Frank, no longer working for NASA or the Air Force, can fly his own planes for his own enjoyment. In summertime, which is fire season, he flies for the people of Montana, taking one of his two planes up to run smoke-spotting missions over the ranches around Yellowstone County. It's a different kind of battle—a local, defensive contest against nature, not an ideological foe. But even at eighty-eight, Borman is happy to volunteer to fight it.

He continues not to feel terribly sentimental about Apollo 8. Still, mostly because people kept insisting that he should, he would occasionally look at the moon and reflect on the fact that he had flown there. He would try to summon up the dreamy feelings everyone said he was supposed to feel. With a little effort, he'd get a flicker of what they were talking about, but it was something of a bother, and after a while he quit giving it much thought.

Like Lovell and Anders, Borman doesn't specifically recall many of the letters and cards he received after the mission to the moon. That is true for most of the Apollo astronauts. There was just too much mail, especially for Apollos 8 and 11, which the NASA public affairs folks had expected, and, later, for Apollo 13, which nobody had predicted.

But Borman does remember one telegram—from a sender he didn't know—and he still likes to talk about it. The telegram said, simply, "Thank you, Apollo 8. You saved 1968." That, Borman realized, made him feel happier than gazing up at the moon ever did.

AUTHOR'S NOTE

THE FLIGHT OF APOLLO 8 WAS AN EXERCISE IN UPLIFT FOR THE nation, and revisiting the first mission to the moon as I wrote this book was very much the same experience for me. It's a story I've wanted to tell for a long time, and I'm thrilled to have had the opportunity to do so after many conversations over the years with a number of the people who made the historic flight possible.

My job was made far easier—and far richer—because of the help of the three men who flew the mission: Colonel Frank Borman, Captain Jim Lovell, and Major General Bill Anders. Borman and Anders did not know me before I began my work, yet they made themselves available for interviews and answered all of my questions thoughtfully and candidly.

The two days I spent in the summer of 2015 visiting Frank Borman in his gleaming hangar at the Billings, Montana, airport where he keeps his two planes were unforgettable experiences. He is exceedingly candid and unexpectedly funny and projects a deep and fundamental decency.

Jim Lovell, unlike the other two men of the Apollo 8 crew, is someone I have known long and well, ever since 1992, when we began working as coauthors and collaborators on the book *Apollo 13*. In the years since, I have been privileged to call not only Jim but the entire Lovell clan friends, and I have always remained mindful of what a gift that is. In the summer of 2012, my wife and daughters and I spent the weekend with the Lovells at their home outside of Chicago; during our stay, Jim took us to Chicago's Museum of Science and Industry to see the Apollo 8 command module. On our way there, I told our girls, who were eleven and nine at the time, "You may not be quite old enough to appreciate it now, but this will be like Columbus showing you the *Santa Maria*." As it turned out, they were old enough to appreciate the experience—and they still do.

I am indebted, as well, to other veterans of the Apollo program who took the time to share their thoughts and reliably answered my calls and e-mails whenever I had questions—which, I confess, was often. Those people, listed alphabetically, are: Michael Collins, Gerry Griffin, Gene Kranz, Glynn Lunney, and Milt Windler. I also benefited from the dozens of interviews I conducted years ago when Jim Lovell and I were writing *Apollo 13*. Some of the people whose recollections again proved helpful were John Aaron, Jerry Bostick, Pete Conrad, Chuck Deiterich, Dick Gordon, Chris Kraft, Sy Liebergot, Jim McDivitt, Wally Schirra, and Guenter Wendt. For *Apollo 8*, I also had the singular experience of interviewing Gene Smith, a close friend of Gerry Griffin's and an Air Force fighter pilot who spent 1,967 days as a prisoner of war in North Vietnam during the years the Apollo program was flying. The material from our conversations—as can often happen—did not make it into the final version of the book, but my appreciation for his time and his service to the country is undiminished.

The canon of space books is deep and long, and a number of the best of those books were helpful to me as I worked. They included: *Countdown*, by Frank Borman; *A Man on the Moon*, by Andrew Chaikin, *Rockets and People*, by Boris Chertok; *Carrying the Fire*, by Michael Collins; *Flight*, by Chris Kraft; *Failure Is Not an Option*, by Gene Kranz; *Genesis*, by Robert Zimmerman; and *Apollo 8*, the official NASA mission report.

Perhaps the best source of material on the history of NASA is NASA itself. The Johnson Space Center History Office, in particular, offers an almost bottomless wealth of documents, mission transcripts, images, and more. Especially valuable is the center's Oral History Project, which makes available more than a thousand detailed interviews with key figures from NASA's past, many of whom are no longer living. It is a gift to history that this project exists.

Throughout this book, all conversations inside the spacecraft and between the astronauts and the ground were drawn from NASA mission transcripts. In some cases, the exchanges were edited or compressed for clarity and readability; in no event was the meaning or context changed. Conversations for which there is no historic transcript were reconstructed from interviews with the principals involved or from biographies and autobiographies.

The confidential exchanges that took place in the Soviet Union's Central Research Institute building, outside of Moscow, were originally reported in volume 4 of Boris Chertok's four-volume *Rockets and People*. The sources for details about Chris Kraft's meeting with Admiral John McCain Jr. and Susan Borman's conversation with Kraft about the odds for Apollo 8's success were interviews with Kraft and Borman included in the PBS American Experience program *Race to the Moon*.

My most readily available archive was the searchable online history of *Time* magazine, which, along with *Life* magazine, was the country's leading source of space news during the days of the Mercury, Gemini, and Apollo programs. Thanks to managing editor Nancy Gibbs for permitting me to use the original reporting files that were telexed and wired from Houston to New York during the Apollo 8 mission—and survive today on perforated printer paper and yellow onionskin. I have been privileged to be a part of the *Time* family for more than twenty years, and could happily do another twenty.

A word of appreciation, too, goes to the *New York Times* for its TimesMachine site, an archive of every issue of the paper going back more than 165 years. It is an extraordinary tool for researching both historic events and, more important, the context in which they unfolded.

I found additional valuable material at the Lyndon Baines Johnson Library and Museum in Austin, Texas. The facility fulfills both parts of its mission—as a library and a museum—extraordinarily well.

In one of life's wonderful bits of serendipity, the original proposal for this book landed in the hands of the talented John Sterling of Henry

Holt and Company. It was John who, in 1992, saw the power and potential in the story of *Apollo 13*, signed me to my first publishing contract, and shaped that book into what it became. And it is John who, two decades later, worked the same magic with *Apollo 8*. His passion for the story and his exacting method of editing were perfectly expressed when he explained to me, "I read hot and I edit cold." It is precisely the balance that every manuscript—and every author—needs. Thanks also to copy editor Bonnie Thompson for bringing a jeweler's eye to the business of reading and correcting the manuscript.

Like every book I've ever written, this one would not have been possible without the wise guidance of Joy Harris of the Joy Harris Literary Agency. Joy, like John, made *Apollo 13* happen so long ago, and she has shared with me all of the literary adventures that have unfolded since. Over time, the qualities of an extraordinary agent and an extraordinary friend have come together in a single person, one I'm deeply lucky to know.

Finally, thanks and love go to my wife, Alejandra, and my daughters, Elisa and Paloma. They have understood, as I've written this book, the music I find in the moon—just as they've helped me see the lyricism in so many other things.

INDEX

ABOUT THE AUTHOR

JEFFREY KLUGER is the author of nine books, including *Apollo 13* (originally published as *Lost Moon*) and *The Sibling Effect*. As a science editor and senior writer for *Time* for more than two decades, he has written more than forty cover stories for the magazine. He lives in New York City.

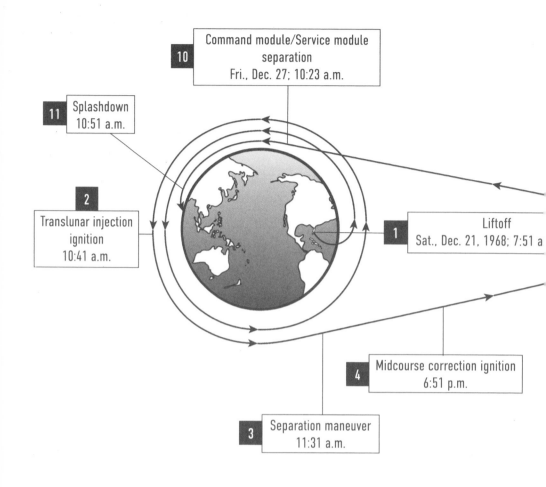

10 Command module/Service module separation
Fri., Dec. 27; 10:23 a.m.

11 Splashdown
10:51 a.m.

2 Translunar injection ignition
10:41 a.m.

1 Liftoff
Sat., Dec. 21, 1968; 7:51 a

4 Midcourse correction ignition
6:51 p.m.

3 Separation maneuver
11:31 a.m.

THE VOYAGE OF

APOLLO 8

Catherine Fitzgerald

Developing Leaders